3/11/91

D0161495

ISLAND AFRICA

The Evolution of Africa's Rare Animals and Plants

Jonathan Kingdon

ISLAND AFRICA

The Evolution of Africa's Rare Animals and Plants

Jonathan Kingdon

PRINCETON UNIVERSITY PRESS
Princeton, New Jersey

First published in Great Britain by William Collins Sons & Co. Ltd
First published in the United States of America in 1989 by
Princeton University Press, 41 William Street, Princeton, New Jersey 08540

Library of Congress Cataloging-in-Publication Data
Kingdon, Jonathan.
Island Africa.

Includes index.
1. Rare animals—Africa—Ecology.
2. Rare plants—Africa—Ecology.
3. Natural history—Africa.   4. Evolution.
5. Nature conservation—Africa.
I. Title.
QL84.6.A1K56   1989   574.5′29′096   89–10579

ISBN 0-691-08560-9

Printed in Great Britain

Contents

Preface

This book had its beginnings with a letter from Dar-es-Salaam in Tanzania, with news of a plan to systematically fell the East Usambara Forest using the latest technology and expertise from Scandinavia. In common with many other naturalists that were acquainted with this beautiful forest and its unique flora and fauna I was horrified. Equally outrageous, this stripping of Tanzania's precious natural assets was masquerading as 'Foreign Aid'.

I shared my dismay with colleagues from the Fauna and Flora Preservation Society and protests from all over the world eventually persuaded the protagonists of this plan to scale it down. The Usambaras are but one of Africa's many centres of endemism where numerous species cluster in a single locality. Some people are aware of the existence of such centres and of threats to their survival but the Usambara crisis emphasized to me the need for a broader assessment of their continental, indeed global significance. The campaign to save the Usambara forest also revealed the importance of single individual initiative as well as concerted social action. One of the rewards of writing this book has been to witness the spirit and strength of ordinary people acting in defence of values that are not purely economic. I have therefore tried to attach contemporary realities and action to a subject with perspectives that span tens, even hundreds, of millions of years.

Some years ago I completed an inventory of East Africa's mammals which was the vehicle for a collection of exploratory essays on evolution, indeed the subtitle was 'an Atlas of Evolution in Africa'. Because evolution is again a central concern, this book could have carried a similar subtitle but this time I have sought to attach some of the practicalities and dilemmas of the contemporary scene. The subject has broadened into a continental natural history, but with a voluntary restriction to rare animals and plants from a dozen or so offshore islands and a similar number of land-based ecological islands. This has given manageable proportions to a vast canvas.

It is obvious that islands and the communities that are isolated in them are appropriate vehicles for essays in evolution but it is no less obvious that finite numbers of rare animals and plants should invite our concern for their survival. A rare thing embodies both the fragility of existence and the uniqueness of all life, so when a single locality supports large numbers of rare species, both our concern and the need for explanations acquire an intense focus. At a time of generalized prescriptions I reassert the uniqueness of places and species. Such affirmation is a necessary counterbalance for that mainstay of scientific thought and practice—the comprehensive explanation.

Explanations tend to be shaped by the conceptual frameworks in which both our predecessors and we ourselves have fitted them. With this in mind the book has provided me with a pretext to re-examine some of what I have learnt. I am also aware that knowledge of Africa and the vocabulary used to describe it owe a lot to historical accident. Mariners named Zanzibar, Cape Verde and the Ivory Coast. Aardvark, hartebeest and fynbos are inherited from the dialects of southern colonists while the maps and categories of early naturalists have left outlines in books and minds no less than empires have left political frontiers. More recently, widely-read naturalists have established categories or theories that may actually obscure our further understanding. The accidents of personal history also limit and shape what is learnt and this book is a personal attempt to reassess the natural history of the continent of my birth. It has grown out of an early interest in rare animals and more recent visits to their habitats—it is the outcome of a succession of journeys and, in the chapters that follow, the islands are described as if they were a series of landmarks seen on a voyage.

Mt Kilimanjaro.
Approaching from above . . .

Landmarks can be seen from many perspectives. I have, for example, flown over and around Mt Kilimanjaro, a very different perspective to seeing its banded valleys from the plains below, or glimpsing its snows from within the forest on its flanks. This book therefore attempts to move easily between three very different perspectives: the lofty generalized overview, then the closer search for pattern in a landscape and finally close-up detail, the fabric of life in a very particular place. Space travel and satellite photography have had a profound influence on our perceptions over the last two decades. Through film and photography many people can now share something of the French astronaut Jean-Loup Chretien's delight in looking down 'and distinguishing without difficulty the little details of the place where I was wandering on foot some weeks earlier . . . some seconds later we were flying above the USSR'. The Russian Vladimir Koralyonok has described seeing an orange cloud formed by a dust storm over the Sahara rise up only to dissipate into a rain storm over the Philippines, while his colleague Aleksei Leonov remarked that he never knew what the word round meant until he saw the Earth from space, so absolutely round is the world!

As one brown patch on a blue globe Africa can now be visualized as a rather dry island where white vapours off the surrounding seas shed their rain rather sparingly. Climates past and present, global and local, have become an ever more insistent part of our perceptions and this too is reflected in the structure of this book, which portrays many of the centres of endemism as coasts with coastal climates.

I do not think that the distant view is incompatible with the appreciation of small things witnessed in our brief moments and on our own scale. In each chapter I have sought to give a 'sense of place', to focus on the things, above all the living things, that most distinguish the islands concerned. Animals and plants *are* places and especially those that have remained a permanent part of the place and the processes from which they evolved.

In spite of a linear progression I hope that other dimensions of space and time will not be lost to view and that a broad and dynamic view of Africa's natural history will emerge from this picture of its enclaves. The book has been written because I thought these treasuries of rare animals and plants, Africa's centres of endemism, have been neglected. It has been written because the boundaries of these centres are narrow, the numbers of organisms that inhabit them are finite, their conservation is an urgent priority and for some centres the hour is late. It has also been written in tribute to the many people, past and present, who love and value these places and their natural inhabitants.

I am indebted to H. Walter for the climatic diagrams that accompany each centre of endemism map (see p. 14). They are drawn from a world climate Atlas (Walter and Lieth, 1960–67) and are the adaptation of a model devised by Gaussen in 1955.

I have had the privilege of much kind help from many quarters during the preparation of this book and over years of travel in Africa. Foremost has been the support of my immediate family, parents, children and Elena Kingdon who, with Zuleika, typed the manuscript. Of the many institutions in Africa that have given me hospitality or support I am especially grateful to Makerere University, Uganda, the National Museums of Kenya, Mweka Wildlife College, Moshi, Tanzania, the Mammal Institute in Pretoria, the Wildlife Department of the Southern Region of Sudan and the Ethiopian Wildlife Department in Addis Ababa. In Europe I have had the support of Oxford University Zoology Department, the Zoological Society of London and the Wellcome Trust.

I am indebted to the staff and for the facilities of the National Museums in Nairobi and Kampala. the British Museum in South Kensington and Tring, Oxford University Museum, the Paris Natural History Museum, Kew Herbarium

its valleys from below . . .

and Library, the Oxford University Libraries, CNRS in Paris, the County Museum Los Angeles, the ZSL Library, the Royal Geographical Society Library, IUCN and ICBP in Cambridge, O.R.S.T.O.M., Adiopodoume and the Wisconsin Regional Primate Research Station, Lake Tissongo, Cameroon. Thanks also to Kyoto University for a visiting professorship to the Primate Research Institute at Inuyama, Japan.

A special debt of gratitude is owed to my host in Oxford, Sir Richard Southwood, and to David Jones in London, to Phil Agland, Francois Bourliere, Tom Butynski, Malcolm and Chris Coe, Andrew and Linda Conroy, Gerard Galat, Steve Gartlan, Jan Gillett, Jean-Pierre and Annie Gautier, Alan Hamilton, Chris and Sheila Hillman, Mike and Leslie Lock, Jon Lovett, Angus MacCrae, John Phillipson, John Skinner, Tom and Lisa Struhsaker and Tim Synnott, all of whom have discussed ideas and drafts or given practical help in a variety of ways.

In letters, papers, discussion and for stimulus and hospitality I am grateful for the help given by many people. Among them I should mention M. Abdi, P. Anadu, E. Ayensu, R. Bailey, F. Bourliere, A. Bandusya, R. Buxton, A. Brosset, J. Butyinski. H. B. S. Cooke, W. and B. Carswell. A. Darwish, I. Deshmukh, B. Dutrillaux, D. Estes, K. Eltringham, V. Funaioli, T. Fison, A. Guillet, M. D. Gwynne, B. Grandison, M. Gosling, R. and J. Glen, J. Green, J. Harris, M. Harrison, D. L. Harrison, P. Heard, C. Hemming, K. Howell, O. Hedburg, R. Hughes, B. Juniper, K. Joycey, C. Jermy, H. Lamprey, I. M. Lernould, R. and M. Leakey, W. May, R. Martin, D. MacDonald, G. Lucas, P. Miller, J. Morris, J. Oates, I. Parker, M. Pickford, D. Pye, D. Pomeroy, M. Rae, G. Rathbun, I. Redmond, A. Root, M. Ruvolo, P. and S. Sandford, A. Schiotz, J. Sabater-Pi, S. Stephenson, L. Silcock, A. Teshome Ashine, P. Williams, F. White and R. Wise.

Finally I must thank Crispin Fisher, Sarah Amit and Myles Archibald at Collins for their friendly cooperation in the production of this book.

from within the forest

Introduction

The Centres of Endemism

To all appearances Africa is one of the most homogeneous of continents and its best-known animal and plant species can be partitioned between three or four major zones that stretch from one shore to another. Desert, savannah, forest and the uplands of east and south—these are the ecological communities or categories on which African natural history has long been based. Plants and animals within these zones or communities have very wide natural ranges and account for about three-quarters of all African species. The remaining quarter are rare or localized species that cluster in recognizable enclaves or 'Centres of Endemism'. It is these enclaves and their inhabitants that are the subject of this book. Because these communities are distinct from the surrounding 'seas' of desert, savannah, forest, the Centres of Endemism can indeed be visualized as land-locked islands. Some have boundaries as crisp as a water-line or cliff face, some stand up like blocks or domes, others can only be thought of as blurred stains on the continent's foreshore, because their distinctness lies more in the composition of their flora and fauna than in any conspicuous feature of their landscape. All are minuscule compared with the vast land mass of which they are a part.

In a snapshot of the contemporary scene some Centres of Endemism will appear peripheral, but their position becomes very much more pivotal and significant when Africa is examined as a theatre of enormous climatic and evolutionary changes. In a continent where climates and habitats have fluctuated wildly, islands of relative stability have conserved many species, and it is their role as natural conservation centres that needs greater recognition. Around each stable centre there were gradients towards greater wetness or dryness, more heat or cold. Wherever these gradients were fragmented or dislocated by broken terrain repeated cycles of climatic expansion and contraction shifted or stranded local populations. These were ideal conditions for the evolution of local species or subspecies within localized pockets. These Centres of Endemism are therefore closely associated with foci for evolution. Islands are also classic refuges for long-range wanderers. Almost all oceanic islands were colonized in this way. On the mainland, such colonists are likely to have arrived and then been stranded during periods of climatic change or at times when Africa came into contact with Eurasia. Their subsequent survival may then have depended upon the persistence of their particular 'slice' of habitat or on their own adaptability. Most Centres of Endemism therefore contain more than one type of endemic. Stability has favoured the conservation of older types (commonly called 'old' or archaeoendemics) while a diversity of habitats in the vicinity has encouraged 'new' or neoendemics.

Geology and Endemism

How old are the oldest endemics? This question can conveniently introduce a brief account of some of the geological forces that have shaped and isolated the continent and influenced its climates, habitats and inhabitants.

When studying the evolution of unique island communities there are three very different time-scales that need to be considered. During the first, geological time, the physical structure of the land has been determined over hundreds of millions of years, the second, climatic time, concerns fluctuations of climate over tens of millions of years (and with it vegetation), while the third, historic time, is very largely the story of changes that have taken place during mankind's rise to ascendancy.

One good reason why the biogeography of animals and plants has tended to be discussed separately is the very different time-scales involved. It is not unusual for an individual plant to live hundreds of years. The rate of genetic change in such a species can be so infinitely slow that it needs millions of years to show up. Thus, many genera of living plants are known from periods before the break-up of the continents.

Although it is known from fossils that some primitive animal types (such as dragonflies, crocodiles and certain birds) differ very little from ancient extinct animal forms, few are as old as the more primitive types of plant.

Animals and plants also differ very greatly in their modes of dispersal and in the mechanisms (and rates) of reproduction and speciation. It is quite likely that wild plant species have arisen through hybridization, something that is very exceptional in animals. It is also likely that a distinctive mutant plant has a better chance of survival as a reproducing individual than an aberrant animal. It has even been suggested that high rates of plant endemism in the Cape, Madagascar and Australia could have been assisted by gene changes due to solar radiation over the southern hemisphere. Such influences, if they occurred, would probably be of no real significance for animals.

Notwithstanding the differences, endemism in Africa seems to have arisen in the same localities among both animals and plants, and generally because of the same factors. It is with good reason that we can speak of endemic communities.

Some endemics, especially in southern Africa and that drifting continental vestige, the Seychelles, are extremely primitive. For example, several of the palm trees and all the caecilian amphibians of Seychelles are completely unable to tolerate salt water and almost certainly never have been able to tolerate it. They have probably survived from the period when both Africa and the Seychelles were integrated in the super-continent of Gondwanaland. Both palms and amphibians are organisms which have changed in only minor details of anatomy since the Jurassic period, some 200 million years ago. The lungfish, *Protopterus*, and bichir, *Polypterus*, both belong to even older lineages of non-marine fish (species of *Protopterus* and *Polypterus* have remained contained within single river-catchment areas in spite of connections and disconnections with other river systems that took place millions of years ago). The only surviving lungfish outside Africa are in South America and Australia.

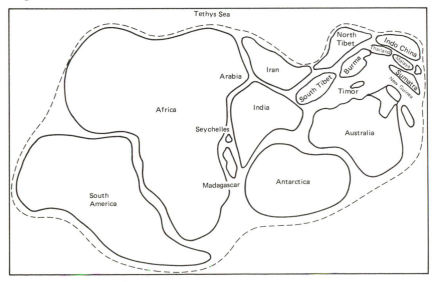

The supercontinent of Gondwanaland, 150 million years ago. Some species, such as some of the palm trees and caecilians on Seychelles, or lung-fish in Africa, Australia and South America have kept their 'Gondwanan' character with relatively little change.

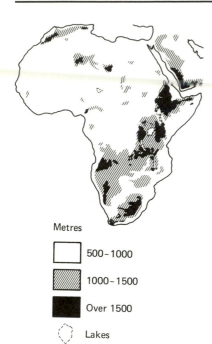

Metres

500 – 1000

1000 – 1500

Over 1500

Lakes

The topography of Africa.

Within Africa, the southern tip has been the principal region where ancient plants and invertebrates have persisted in little-changed forms. Some have their closest affinities with Gondwana fossils and contemporary species in refuge areas of South America and Australia.

Africa is the largest fragment of Gondwanaland. After nearly 200 million years of relative stability, perturbations deep within the earth first dimpled and then tore this land-mass apart between 270 and 200 million years ago. Even to-day, the blunt contours of this great continent carry the record of its geological history, and nowhere is the tenuous grip of life on its corroded surface more evident than in satellite photographs, where vegetation shows up as a faint, uneven stain.

The continent's major aspect is of vast alluvial basins and plains textured by the endless branching of streams. There are the great living rivers, the Nile, Zaïre, Niger and Zambezi, yet many more are dead, dry drainage lines brushed far and wide over expanses of sand and rock. Slashed down the eastern side of the continent is the Great Rift Valley which on satellite photographs looks like a badly healed wound; the great lakes puddle in its scars, their liquid darkness reflecting ultramarine blue at the cameras in space. Further north another valley has split open to let the ocean in—the Red Sea.

These tortured surfaces are eloquent of the upheavals and deep global currents that tore Gondwanaland into fragments. Few of the old Gondwana surfaces have survived hundreds of millions of years of tilting, fracturing, erosion and burial beneath volcanics and sediments; only on raised plateaux with flat tops do they escape being overlaid. Rarer still does an ancient landscape

Africa's rivers and lakes.

occur in an area of clement and stable climate. Blocks of ancient rocks have been thrust up and then cut back. Among these are massifs in the Sahara, rift-edge mountains between Ruwenzori and Lake Malawi and the crystalline mountains of eastern Tanzania. The chains of ancient mountains run on south, reaching over 3480 m in height in the Drakensberg and ending up in the eroded landscapes of the Karroo and the Cape. The oldest Centres of Endemism are associated with these mountains.

By 100 million years ago all the continental masses of Gondwanaland, including Madagascar, were separate, and Africa (of which Arabia was still an integral part) had begun a very long period of total isolation. At about this time, Africa's last connection with South America was broken (yet in both continents the fresh-water rivulin fish, Cyprinodonts, have retained many similarities and have been prone to local speciation).

All mammals and effectively all birds postdate the break-up of Gondwanaland and their exchange between continents and islands must have been due to flight, rafting, or the collision of land-masses. Even ancient reptiles, such as tortoises (which are particularly prone to endemism in the Cape and on oceanic islands) have established themselves less than 50 million years ago. While some other continents drifted extensively, Africa maintained a sufficiently stable position on the globe for the southern end to retain a relatively stable climate and thus conserve many very primitive species. Many millions of years of isolation were broken when Africa touched Eurasia about 30 million years ago. Numerous mammal groups flooded in and out of Africa at this time. The anthropoid primates and the elephant's ancestors migrated out for the first time while even- and odd-toed ungulates (ancestors of today's bovids and horses) as well as hares, modern rodents and carnivores all made their first entry into Africa. After a long period of conjunction with Eurasia there was a further break, which may have lasted some 6 million years until contact was regained about 17 million years ago when another influx and emigration took place.

Such intercontinental exchanges have left their mark on endemic communities. For example, advanced murid rats entered Africa during the mid-Miocene connection. In equatorial regions they proliferated into many species and displaced older forms of rodents. The latter only persisted in the more difficult environments where they had a head start (such as gerbils in deserts) or became specialists (such as cricetid climbing mice).

In the extreme south of Africa, murids were slower to displace older types of rodent. As the ornithologist James Chapin remarked for birds in 1932, the earlier inhabitants of a country have many advantages in competition for an established territory, but any extreme change in vegetation or the arrival of highly adaptable immigrants is likely to swing the balance against them. Over most of Africa both these forces were frequently repeated hazards.

Geological upheavals have created physical islands where montane habitats and communities developed. They have also generated barriers and redistributed moisture by capturing rain on one side and starving nearby lands of rain on the other. All down Africa great mountain chains are associated with rifting. The rift valleys have derived from several phases of uplift, the most recent being 22, 6 and 2½ million years ago. The brittle crust arched up in domes, the largest in Ethiopia being 400 km wide. Aligned over thousands of miles in a jagged course from north to south, the uplifted face has split along its axes to form rift valleys that average about 40 km in width. These vast upheavals not only lifted up plateaux, but associated volcanic eruptions have spewed out over extensive areas and thrown up volcanoes such as Kilimanjaro, which is

North-east Africa/Arabia over ten million years ago, showing the formation of domes which split open to form the Rift Valleys and Lakes, the Red Sea and Gulf of Aden.

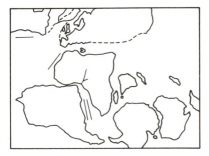

The shape of Africa 100 million years ago. Populations of Rivulin fish (Cyprinodonts) in Africa and South America may have separated at about this time. Primitive insectivores may have crossed over the narrow channel between Africa and Madagascar some time after this.

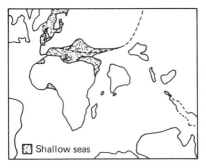

The shape of Africa 50 million years ago. The distinctive character of African fauna and flora developed during a very long period of isolation before Africa and Eurasia collided about thirty million years ago.

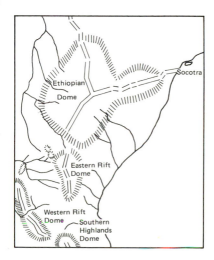

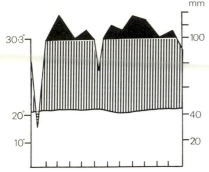

The climate diagrams used throughout this book are derived from Walter and Lieth (1960–67) and Gaussen (1955) and allow an immediate visual assessment of climates within a region, as well as allowing comparison between regions. This is accomplished by plotting mean monthly temperature and rainfall on the same scale (20 mm of rain = 10°C). A period of relative aridity occurs when the rainfall curve falls below the temperature curve (the stippled area), whilst a humid period occurs when the rainfall curve rises above the temperature curve (vertical hatching). Very high rainfall is plotted on a reduced scale (black area).

5895 m high. All the earlier volcanoes have eroded away and deposited sediments that are thousands of metres deep. Cycles of erosion, deposition and the cutting back of mountain sides and gullies have followed each of these events, leaving a complex and often dramatically beautiful landscape. In the rain-shadow of these mountains dry tracts have formed, which add climatic and ecological barriers to the physical ones of mountains and rivers.

Relatively recent volcanics and tipping of land surfaces have changed the direction of rivers and diverted waters and their faunas from one basin to another. For example, the Kagera River flowed westwards towards the Atlantic until volcanic eruptions blocked its way and reversed its flow eastwards into Lake Victoria. Likewise, volcanoes blocked Lake Kivu's drainage into the Nile system and forced it to find an outlet to the south into Lake Tanganyika and hence into the Zaïre.

Each of these events has been associated with the isolation or diversion of animal populations and the evolution of new forms. It is easy to think of these as prehistoric events which have ended with the shape of the land as we know it now. In fact, these geological processes are in full swing, even today, just as the evolution of animals and plants is also an active contemporary process, albeit an infinitely slow one.

Climatic Change and Endemism

The geological history of Africa is overlaid by its climatic history. The repeated drying out of moist habitats followed by their renewal and expansion has led to repeated isolations and driven entire ecosystems to adapt to the changes.

Warm seas have been the ultimate sources of rainfall and moist habitats in Africa. Wherever and whenever the seas were too cold for water to evaporate, a continent already prone to drought was still further starved of moisture. Much of Africa's coastline has shared fluctuations in rainfall and temperature with its interior. However, there have been stretches of coast that were consistently moist or dry and these can be identified as relatively stable and predictable foci

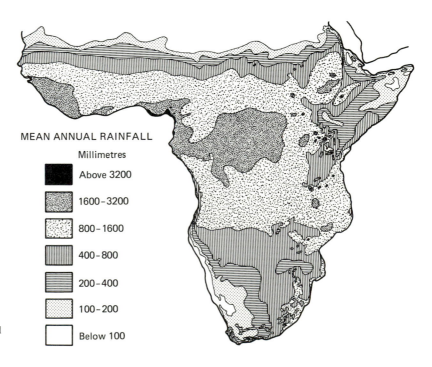

MEAN ANNUAL RAINFALL

Millimetres

- Above 3200
- 1600–3200
- 800–1600
- 400–800
- 200–400
- 100–200
- Below 100

The contemporary mean annual rainfall in Africa.

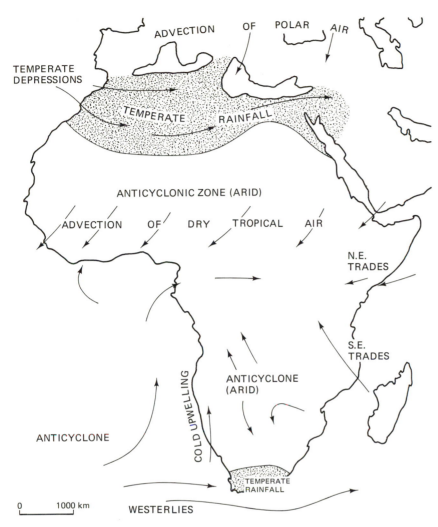

How the atmosphere circulated over Africa during the last glacial maximum (about 18,000 years ago). Modified after A. Hamilton.

in a continent subject to unstable climates. The interior has had its own inland coasts. There are some mountains where clouds blowing in from the sea have always shed their rain while others have not escaped periods of drought. Because they are the primary interfaces between dry land and the life-giving influence of water these strategically placed coasts and mountains have provided most of the chapter headings for this book.

Many endemic species may have evolved in much less restricted theatres, but the places in which they have accumulated and where we find them today are immediately identifiable with a particular coast or mountain chain.

Dry climates have been much more extensive than moist ones, and more than once forests have shrunk back to form minuscule refuges in moist pockets. As a result the great majority of contemporary forest species show signs of being derived from more arid-adapted forms. For example, many arboreal species, notably hyraxes, squirrels and monkeys, derive from terrestrial or semiterrestrial ancestors. The timing, duration and extent of global climatic changes are still very imperfectly known but the chart overleaf summarizes the current state of knowledge.

Over a period of 15 million years or more, but especially during the last million, the world's climate has repeatedly swung from wet to dry and from warm to cool. Several times forests have spanned the equator, broad and wide all

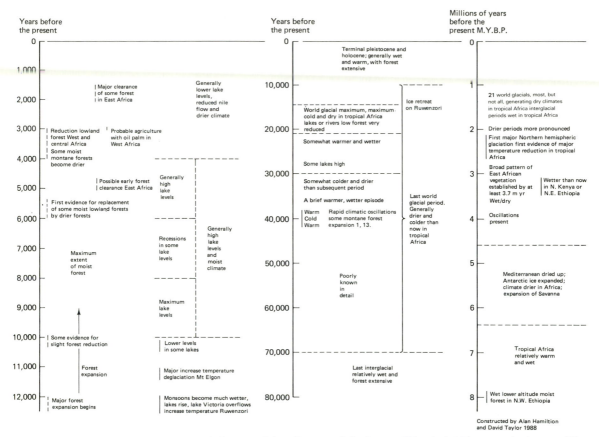

Years before the present				Years before the present		Millions of years before the present M.Y.B.P.	

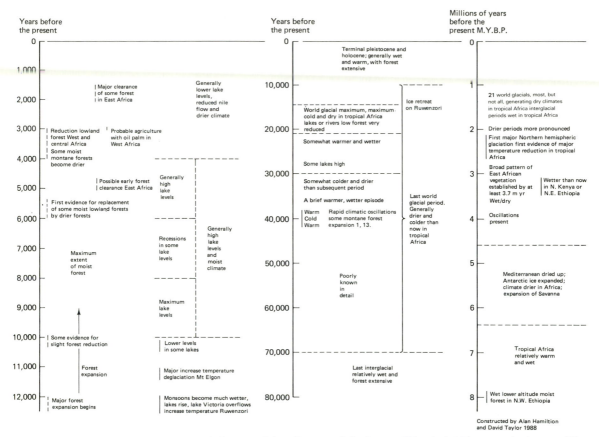

The history of the climate and the forest of Tropical Africa over the last 8 million years, constructed by A. Hamilton and D. Taylor.

Constructed by Alan Hamilton and David Taylor 1988

Zones where major biomes (ie desert or forest) are likely to have interacted during periods of climatic change (mostly along consistent and relatively narrow interfaces).

the way from Guinea to Mt Kenya, only to be succeeded, as a result of droughts, by deserts that sometimes linked the Sahara with the Kalahari. For much of this period and earlier still the continent has been very dry, and today's climates are probably closer to the moist, warm end of the scale. A map of the extent of wind-blown sands in Africa is a reminder of how very dry Africa has been (see next page). In the face of such unpredictable climates the majority of Africa's 'staple' species have had to become exceptionally adaptable and their wide ranges are a reflection of this. Many other species have ebbed and flowed with the tides of climate, especially the vulnerable forest species which can live nowhere else.

Yet, amid all the changes, there have been identifiable areas with stable climates. Somalia has always been a hot, dry spot, Namibia a cold one, Cameroon has been consistently warm and wet, Ethiopia high and dry, Ruwenzori high and wet. The coast and mountains of equatorial east Africa have always caught rain from the Indian Ocean. Of equal significance have been the boundary zones where major variables of the environment meet consistently in the same area. Wet meets dry, high meets low, cold meets warm. In these narrow zones, which are almost always on hills, mountains or escarpments, there is a great variety of biological niches because so many permutations are possible. South or north faces tilt towards the rain or the rain-shadow, towards moist or desiccating winds. There are high valleys and low ones, rich soils and deficient soils, alkaline gravels and acid sands, clays and peats. There are only relatively few places, how-

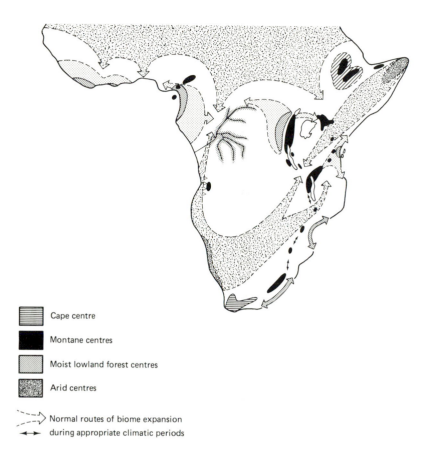

Cape centre

Montane centres

Moist lowland forest centres

Arid centres

Normal routes of biome expansion
during appropriate climatic periods

Centres of endemism and their interactions. Key shows Cape centre, montane centres, moist lowland forest centres and arid centres. Arrows show normal routes of biome expansion during appropriate climatic periods.

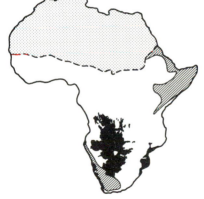

Kalahari sands in southern Africa

The Sahara & southern boundary of fixed dunes

Contemporary arid areas in S.W. & N.E. Africa

Evidence for extreme aridity in Africa. Wind-blown sands extend from the Sahara to form fixed dunes well south of their present limits. To the south, Kalahari sands underlie contemporary forest.

ever, where such climatic boundaries are likely to have been stable over millions of years, but in those rare spots there are many unique animals and plants. In such localities animals or plants could escape any minor change by shifting a few kilometres up or down some valley or further along some coastal plain.

The steepest interfaces between dry and wet climates occur in eastern and southern Africa, where moist coasts are close to arid interiors. Wherever mountains are interposed along the boundary between these extremes, local endemics become abundant. This abundance of unique species is partly due to local adaptation, but how specialized and different the species have become is influenced by how long the environment has been stable. Newly erupted mountains like Kilimanjaro have many fewer endemic species than old mountains such as the nearby Usambaras.

The number of endemics is therefore partly a matter of accumulation (species are less likely to become extinct in a stable environment) and partly of local

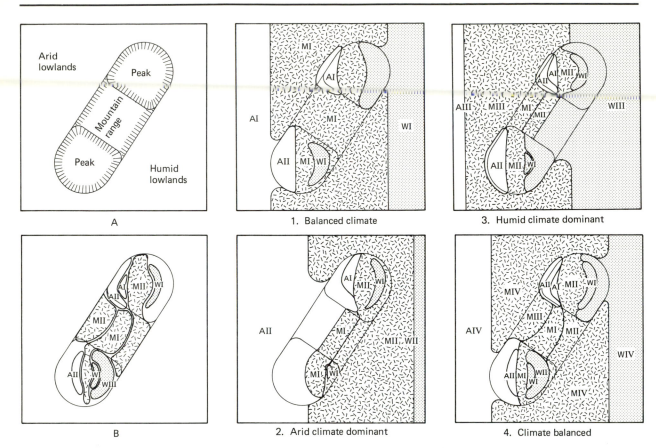

Schematic diagram of a mountain range acting as a 'population trap' and focus for island speciation.
a) A mountain range with two peaks positioned oblique to the dominant climatic influences, aridity from the left, humidity from the right. b) The distribution of seven 'montane endemics' belonging to three lineages. Although imaginary and schematic the pattern is typical in showing species 'stacked' in incomplete altitude bands. Species belonging to the A lineage inhabit the rain shadow and are of Arid lowland derivation. W species are limited to the moister slopes and derive from wet lowland habitats. The 3 M species derive from intermediate lowlands. c–f) illustrate climate oscillations in which three originally uniform populations belonging to three habitats become isolated and differentiate on the mountain slopes. (Compare this scheme with that shown on p. 182.)

adaptation. The first needs time while the second is strongly influenced by how extreme an island climate is. On very high mountain tops with scorching days and freezing nights selection is severe and its effects rapid for the small number of species that such habitats can support (giant lobelias and groundsel as well as some Afro-alpine rodents and insects are thought to have evolved at particularly fast rates). It is the absence or moderation of exterminating climates that favours an *accumulation* of many species. Such accumulation has been most favoured in highly diversified or stratified landscapes, where the habitat may be more thoroughly and subtly partitioned between potentially competitive species (the sunbirds of Ruwenzori are a typical example of numerous species being stacked into a relatively small area; their abundance is also influenced by the central position of these mountains and perhaps by their great age).

Vegetation and Endemism

Consider now the effect of climate on vegetation. The bold outlines of Africa's vegetation are obvious enough and well known. At one extreme a broad equatorial band of forest, at the other, desert in the Sahara, Somali Horn and along the south-western coastline. In West Africa, woodland and steppe form long latitudinal belts, but in eastern and southern Africa these bands are slewed by gradients in elevation as well as latitude. Winds bring in moisture from warm seas (or drought from cold waters), there are rain-trapping uplands and rain-shadows behind them. Peculiar soils further dislocate vegetation belts and generate various unique and localized ecotypes. Almost everywhere cultivation, burning and pastoralism have modified or obscured the outlines and character of natural vegetation types.

A comprehensive map of Africa's natural vegetation was first put together by a small international committee (AETFAT) and published by UNESCO in 1959. This map has provided the baseline for innumerable ecological studies since then. Its principal author, the botanist Frank White, updated and greatly expanded this map in 1983 and a simplified version is shown below. Recent droughts have demonstrated how extensive are the ecological effects of relatively small irregularities of climate pattern. Multiply these fluctuations into the large-scale climatic cycles that are known to have punctuated the last few million years and it is obvious that there have been vast changes in vegetation. What was Africa like in those arid or moist extremes?

In order to visualize the possible extent of changes wrought by extremes of fluctuating climates in the past, the palaeontologist Basil Cooke has taken rainfall maps and drawn isohyets based on 150% and 50% of the present pattern.

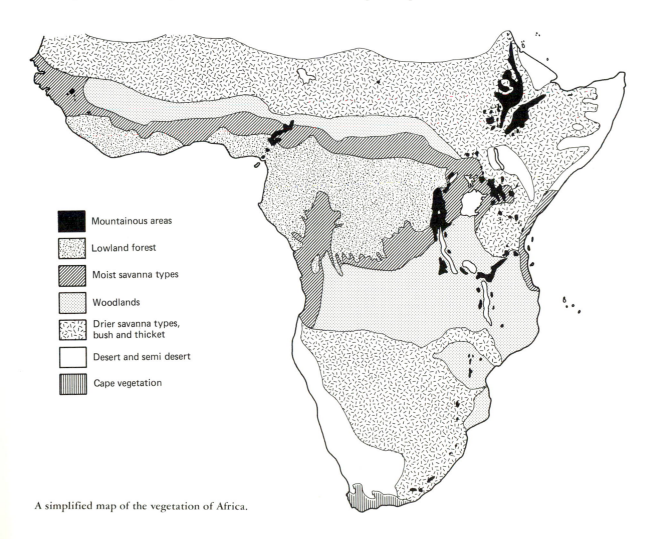

Mountainous areas

Lowland forest

Moist savanna types

Woodlands

Drier savanna types, bush and thicket

Desert and semi desert

Cape vegetation

A simplified map of the vegetation of Africa.

Superimposing these on the southern region of the UNESCO map, he has redrawn the boundaries of each vegetation type. The results are not reconstructions of actual vegetation but they do indicate the kind of changes that would have occurred at 150% and 50% of present rainfall (see next page). In periods wetter than the present, arid-adapted communities would have been separated north and south of the equator by a very broad belt of forest. In dry periods, the forests retreated and long north–south corridors of savannah and semi-desert would have cut right through. Add physical barriers to a succession of climatic extremes and the basis for repetitive isolations, both physical and ecological, can be found all over the continent.

Old and New Endemics

Enclaves therefore need to be understood within the broader history of Africa's past and they illustrate many important aspects of the evolutionary process. It has already been pointed out that enclaves shelter fundamentally different types of endemics. The relict species, or 'archaeoendemics', seldom have close relatives in adjacent areas, although some, such as caecilians or golden moles, may speciate within a very limited region. 'Neoendemics' commonly have close relatives in the same area or in neighbouring habitats. Such species tend to continue to diverge in adaptation to their small new world. A community such as the Cape flora has many such specialists and these species in turn create new opportunities for still further specialization in those animals and plants that come to depend upon them for food and shelter. To some extent, 'old' and 'new' are relative terms. A primitive arthropod is more likely to be an archaeoendemic than a cichlid fish, but within a late radiation of cichlids in a 'new' lake, some will be more recently evolved than others and distinctions can become blurred.

The link between old and new endemics and the age of their island is clearest on oceanic islands. Like their continental equivalents, these may be very young volcanoes or atolls, such as Aldabra, or very ancient, like the Seychelles. They also occur within zones where the climate may be relatively consistent—for example, Tristan da Cunha is always wet and cold, Socotra dry and hot, and Pemba moist and warm—or the climate may have fluctuated. The Seychelles betray the latter in having a forest community dominating the lowlands but several arid-adapted plants and animals surviving as relics on waterless exposed hilltops, one of them the unique Jelly-fish Plant, *Medusagyne*.

Arrivals that are quick to adapt to the peculiarities of an island often alter in size, colour or structure to become distinctive endemics. Certain groups have consistently provided rapidly adapting colonists, notably pigeons, rails, white-eyes and scops owls. In general, these bird types tolerate a very broad range of climates and vegetation, but genetically isolated island populations quickly take on a distinct appearance; this may be emphasized by a natural mutability in their colouring, which leads to the fixing of certain peculiarities of the founders. On new islands all endemics are of this type. Older islands may conserve isolates of types that have meanwhile become extinct on the mainland—some Socotra plants and reptiles appear to belong to this class. Only the oldest and most stable islands conserve truly archaic forms; for the most part these survive best in continental enclaves.

Speciation in Centres of Endemism

The evolution of several endemic species from a single ancestor on a single island is another topic touched upon in this book. On the steep slopes of a lonely mid-Atlantic volcano, very closely related and uniquely St Helenan beetles now occupy circular bands separated only by altitude. In the African lakes cichlid fish

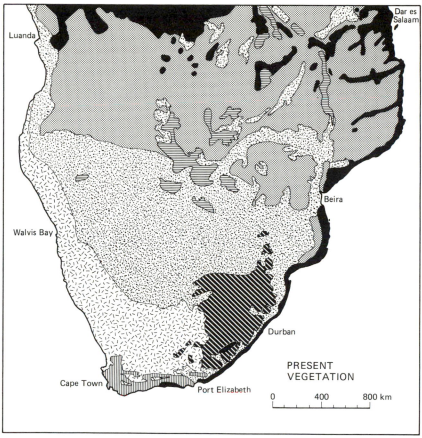

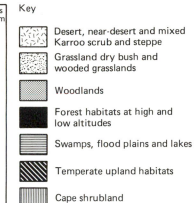

Key

Desert, near-desert and mixed Karroo scrub and steppe

Grassland dry bush and wooded grasslands

Woodlands

Forest habitats at high and low altitudes

Swamps, flood plains and lakes

Temperate upland habitats

Cape shrubland

The effect of a variation in rainfall on Africa's vegetation.
(simplified and modified after H. B. S. Cooke, 1964).

PRESENT VEGETATION

0   400   800 km

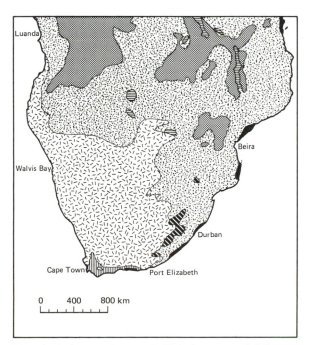

Hypothetical vegetation at 50–60% of present rainfall.

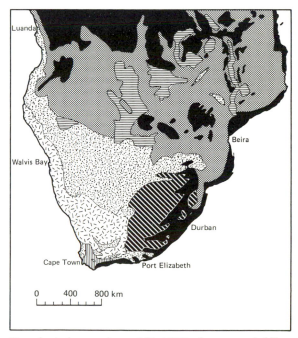

Hypothetical vegetation at 140–150% of present rainfall.

have diversified and multiplied in species, often in the absence of physical barriers. In the immense saucer of central Africa different species of monkey appear to have developed from closely related subpopulations that originated from climatically distinct subdivisions of the Zaïre basin (see plate on p. 197.)

Wherever such animals and plants had become sensitive to the exact balance of conditions in their environment subpopulations would have arisen. These must have fluctuated in size with the ebb and flow of their own particular milieu. The many great changes in climate that are known to have affected Africa would have been decisive in favouring or inhibiting the spread of particular local populations. Periodic isolation would have encouraged speciation but climatic instability would also have favoured species with broad tolerances. In this way, Africa's fauna and flora have tended to diverge into at least two major divisions—widely distributed, common and adaptable species, and more restricted species living in narrowly defined habitats or geographically small regions. These categories need some qualification because most Centres of Endemism have very narrowly restricted species at their heart, while species with more extensive ranges occur on the peripheries. (The map of Cape proteas on p. 54 illustrates this well.)

The Example of Jewel-Weeds

In a book about the rarer animals and plants of Africa one question immediately arises—why are they rare? A stock answer is that human activity has brought them to this pass. For some this may be true, but the majority of species discussed in this book are intrinsically rare or localized for reasons associated with their own biology and history; competition with or predation by other organisms; or adaptation to very localized conditions. Few plants are more eloquent indicators of this and of evolution on a continent of islands than the balsams or jewel-weeds *Impatiens*. Familiar to gardeners as the pot-plant 'Busy Lizzie' (*Impatiens walleriana*), balsams are flimsy leaved herbs which only grow in wet and shady places. Some species are more resistant to exposure than others but none can withstand sustained desiccation or direct sunlight. As a consequence, balsams only grow beside permanent watercourses, on wet mountains or in moist forests. Although there are more than 100 species known from Africa, the great majority of these are localized and this is undoubtedly a reflection of the fluctuating climatic history of a predominantly dry continent, the inherent patchiness of balsam habitats, the tendency for balsams to vary and their poor capacity to disperse.

Although immediately recognizable as balsams and very uniform in their basic characteristics, the few widely distributed types such as Irving's balsam, *Impatiens irvingi*, are prone to regional variation in the shape, size and colour of their flowers. This has important evolutionary consequences because it influences the success of pollination and leads, as in orchids, to specialization in reproduction and hence to speciation.

Balsams offer their pollinators nectar in a sac or spur and attract them with brightly coloured petals. The surfaces of the lower petals are simultaneously display boards and landing platforms, while the upper petal often forms a hood over the stamens and stigma and usually encases the other petals while in the bud. Whether a bee, butterfly, moth or bird will be the most effective pollinator depends on the exact arrangement and shape of the flower's petals, sepals and reproductive parts. For example, on the island of São Tomé in the Gulf of Guinea two endemic montane flowers have quite different proportions. One species, appropriately called the Trumpet Balsam, *Impatiens buccinalis*, has small petals and a large, pendulous orange sepal, which presumably holds enough nectar to reward a larger-sized pollinator. The other species, *I. thomensis*, has large scarlet petals and a short sepal with a narrow spur that mimics the curve of a small moth's proboscis. In spite of such differences in the colour, size and shape of their flowers, other characteristics show that both the São Tomé flowers belong to the same subgroup of balsams represented elsewhere, at high altitudes, in Cameroon, Ruwenzori and east Africa. Such restricted distributions are thought to reflect a very early establishment by balsams in environments that have remained consistently favourable for them (and many other organisms). In the case of São Tomé (which erupted out of the sea more than 10 million years ago) two separate immigrations, both quite early, are implied. The ancient nature of such colonization has further support from east African balsams. The Uluguru Balsam, *I. ulugurensis*, has a similar cloud forest habitat and belongs to the same high altitude group. Yet this flower, one of twenty Uluguru endemics, appears to be closer to Malagasy and south Indian balsams than it is to the other Uluguru species. African balsams belonging to other subgroups also appear to share or have retained elements in common with Asian and Malagasy species (see plate on p. 25).

These connections show that balsam diversity is not simply the outcome of recently evolved species, although this picture of typical neoendemics may be true for some species. Dr Grey Wilson, the principal authority on these flowers, has suggested that these very poor dispersers may have already differentiated into several distinct lineages at the time they entered south-east Asia from Africa, possibly 45 million years ago. Within Africa, balsams have found very different niches in different localities. Thus highland grassland types centre on the Tanzanian southern highlands, epiphitic orchid-like forms in forested west Africa, a series of altitudinal and soil specialists in Uluguru. The superabundance of species in the Ulugurus, belonging to at least seven of the major balsam subgroups, suggests that these mountains may have had a longer history of continuously moist habitats than almost any other area in tropical Africa; a history covering several tens of millions of years.

The 15 flowers shown in the colour plate are very restricted but each belongs to a subgroup with a more extensive range. The pattern of islands across the equatorial tropics and along the mountainous east illustrate how fragmented moist Africa is, yet the pattern differs for each species group. The most localized species occupy those Centres of Endemism that are described in the following pages. They are a reminder of how important it is to view endemism in a continental and even global perspective.

Some more widely distributed balsams may well be newer, advanced or specialized types. Some more restricted species may be less advanced. The confined existence of the latter within a larger zone suggests a pattern which may be significant not only for these flowers but for numerous other endemics as well. The primary minimal focus is foremost a conservation area, but around it lies a larger zone which represents that centre's potential for expansion. The 'peripheral endemics' may be markers for the recurrent expansions and contractions that are a part of every centre's history. Balsam distribution patterns may therefore offer a graphic illustration of the dynamics of evolution, evidence for flux and change in 'Island Africa'.

Exploration Past and Present

Each profile of a Centre of Endemism begins with its geographic position. In many instances I have adopted an archaic name for the region or for the chapter heading. There are several reasons for this: one is to escape the mental boundaries imposed by contemporary nationalism. My assertion that the flora and fauna of Africa are still largely unexplored is also implied in the names used by early mariners and traders for an unexplored continent. The names of commodities attached to places are also reminders that this was where foreign interest in Africa began—natural history in Africa has often been pursued as an adjunct subsidiary to that mercantile history. Most of the commodities that were once precious or pre-eminent in trade are now obsolete; to resuscitate these names is to be reminded of how ephemeral values are. The continuity of natural communities and the great age of some Centres of Endemism stand in stark contrast to our destructive chase after the commodities they produce or once produced, whether ivory, incense or oil.

The profiles continue with a summary of the region's broadest characteristics, its geological and climatic background and what can be inferred about its past. I have not followed any one biogeographic classification because the zones that hold for a plant or a butterfly are not necessarily very close to those that obtain for a mammal or a bird.

Portraits of Communities

For each Centre of Endemism some outstanding feature, or features, of the ecological community is portrayed by referring to the biology of some typical species. Thus, the bustards have a special mention in Namibia and the Karroo, proteas in the Cape, thrushes in Zanj and monkeys in the Zaïre basin. The distribution of birds, such as the turacos, francolins or sunbirds illustrates how like islands are the great continental mountains. Unlike previous authors I have not been preoccupied with a search for montane corridors between mountain isolates, but find in mountains natural refuges for formerly dominant species that are now in retreat or in the process of becoming isolated specialists. If 'montane corridors' across intervening lowlands were real one might expect the occasional montane species to have colonized such lowlands; in fact, colonization seems to have proceeded in the opposite direction with few if any exceptions. At the risk of too subjective a portrayal I have chosen a few species to typify the unique interest of each island and tried to avoid the tedium of reiterating long inventories of categories and species. Abbreviated lists of endemics are given in Appendix I.

It is often assumed that the term 'Centre of Endemism' is synonymous with 'centre of origin' for the more abundant types of endemics. This may be true in as much as, say, daisy-trees developed on St Helena or Saintpaulias in Zanj; but there is ample evidence that such radiations often followed the arrival of a 'pre-adapted' colonist with quite distant origins.

Antelopes, murid rodents and squirrels were Miocene Asiatic invaders, yet they have proliferated immensely in Africa. Cichlids are river fish that have exploded into hundreds of species in the lakes because there were no other fish that could match their adaptability and mutability. Proteas, heaths and sedges have seen a similar evolutionary explosion in southern Africa, partly because

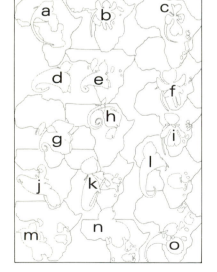

Facing page Some endemic Balsam flowers (*Impatiens*) with their distribution and the total range of the species-group to which each belongs. a *Impatiens nzoana*, b *I. ethiopica*, c *I. rapidothrix*, d *I. buccinalis*, e *I. runssorensis*, f *I. ioides*, g *I. thomensis*, h *I. warbariana*, i *I. cinnebarina*, j *I. macro tera*, k *I. flammea*, l *I. ukagurensis*, m *I. glandulisepala*, n *I. rosulata*, o *I. flanagani*.

Overleaf. Top right, *middle and* lower right, the Dodo *Didus ineptus*, from Mauritius showing details of the head and birds in thin and fat phases. *Top left and* bottom left, the Solitaire *Pezophaps solitaria*, from Rodriguez. The three heads in the centre are of the mainland Green Pigeon, *Treron calva* left, Sao Tome Green Pigeon, *Treron australis saothomae*, and the Tooth-billed Dove *Didunculus strigirostris, right.*

SPOTTED CRAKE   TRISTAN RAIL

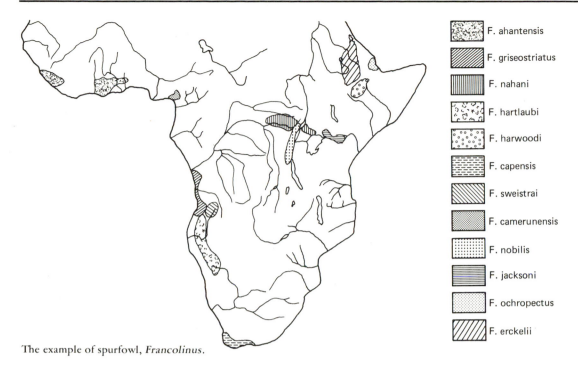

F. ahantensis

F. griseostriatus

F. nahani

F. hartlaubi

F. harwoodi

F. capensis

F. sweistrai

F. camerunensis

F. nobilis

F. jacksoni

F. ochropectus

F. erckelii

The example of spurfowl, *Francolinus*.

they were the plant types best able to cope with the Cape's climate and soils, and partly because the Cape has one of the most finely dissected landscapes in Africa.

Form and Function

In the process of adapting, many such species are radically transformed in appearance. Interpreting such a transformation of an animal or plant type is one of the most fascinating aspects of biogeography. This fascination is one of the reasons for beginning the story of islands with the dodo.

When endemic species change their proportions, elaborate their plumes, brighten their colours or geometricize their patterns, these are the external signs of louder and more specific communication systems, a more crowded community, a narrower niche or a more difficult food. In this book such colours and shapes may serve to embellish pages but it should never be forgotten that these tints and structures are integral to what an organism is, a part of its whole evolutionary response to living in a changing and dynamic world.

Previous page, top. The Zanzibar and other leopards. *Bottom right*, typical East African leopard. *Top right*, Zanzibar leopard with subdivided rosettes. *Top left*, an aberrant mainland leopard in which rosettes have disintegrated into fine irregular freckling. *Bottom left*, aberrant mainland leopard in which normal rosettes have amalgamated into extra large jaguar-like pattern.

Previous page, bottom. A schematic diagram of the inaccessible or Tristan Rail, *Atlantisia rogersi*, and Spotted Crake, *Porzana porzana*, showing the differences in the proportion of flighted and flightless rails. Pink shows the pectoral flight muscles.

Opposite page. Seychelles endemics. *Centre*, Coco-de-Mer palm, *Lodoicea maldivica. Right*, stilt palm, *Verschaffeltia splendida. Top left*, Seychelles Wild Vanilla. *Lower left*, Wright's gardenia, *Rothmannia annae. Lower right*, Jelly-fish plant, *Medusagyne oppositifolia.*

The Enthusiasts

The colour and strangeness of Africa's endemics has long attracted collectors; flowers for gardens and window-boxes, parrots for cages, butterflies for cabinets and tropical fish for aquaria. When the species is rare it is even more sought after and collectors still pose a risk for some species. However, collectomania also has within it the potential for progress. Collectors are now able to learn more, through books and television, about the natural history in the wild of the objects of their passion. Interest is growing in the ecological background of rare animals and plants and there are new opportunities to visit Africa and see such places and habitats at first hand.

With that interest there is growing anxiety that the ecosystems that sustain such exquisite creatures should continue to survive. It was just such a collector, Bernhard Grzimek, whose concern contributed to the saving and subsequent fame of the Serengeti, and there are other less celebrated examples of fanciers or enthusiasts contributing to the protection of particular places far from their own urban backyard.

Partnerships

It is a major purpose of this book to advertise the existence of Africa's Centres of Endemism and to remind every person who collects, admires or simply cares for these rare and beautiful organisms that they have responsibilities as well as opportunities to do something. I suggest that the primary need is for the opening of friendly dialogues, partnerships and a sharing of knowledge and enthusiasm with the people, especially the poorer inhabitants who live in or near these centres. Conservation here is primarily a social, political and human problem. More specifically, it poses problems in communication, in education and in values, because there are huge dislocations in understanding. Starting from the bottom, links have to be made between the various tiers of rural communities, old and new, national citizenries and an external public that wants to help reconcile conservation with development.

Uniqueness

In the following pages I hope to show that rare animals and plants are not the preserve of eccentric specialists, and that the study of their habitats is not a trivial pursuit. I also hope to explore something of the nature of uniqueness. At times it is difficult not to use the vocabulary and form of science, but I hope to assert that the parameters that scientists work within are not the only ones relevant to this issue.

The uniqueness of a rare species and the ecological community to which it belongs can only be experienced in a particular place within very limited boundaries. It is an experience that has been made possible by events and processes that have operated in that place, and no other, over millions of years. That realization is a relatively new one and with it comes the knowledge that we are only just beginning to explore our planet's uniqueness.

CHAPTER 1

Oceanic Islands

The Dodo and the Solitaire

Had the Dodo never existed it might very well have been invented by a cartoonist. Tiny wings and a gigantic beak are comical tokens of a birdhood that has forfeited all claims to grace, yet there is something that has endeared this caricature to generations of people. There are only a few pictures and bits of skin and bone to go by, but an image persists of extreme helplessness. It is as though some unique personality is mourned as much as the extinction of the species Linnaeus named *Didus ineptus*.

At first sight the Dodo appears to be one of the more frivolous caprices of nature—yet its very oddity provides a telling illustration of evolution at work on islands, whether these are oceanic or landlocked.

The Dodo's history becomes slightly less bizarre when it is appreciated that doves, close relatives of its ancestor, are among the bird types that are more consistently stranded on remote islands in all parts of the world's oceans. For a demonstration that the colonization of islands is not pure chance we have only to compare the resident land-bird faunas of two islands, each about 280 km from Africa, São Tomé in the Atlantic and Anjouan in the Comoros group in the Indian Ocean. In spite of being 4500 km apart in different latitudes and oceans they have been colonized by a rather similar selection of some 20 bird families (from a potential of about 70) and of all these birds pigeons are the most numerous and effective colonists.

Resident land birds of two oceanic islands (of similar size and distance) off Africa. The similarities suggest two levels of selection, one for bird species prone to landfall on oceanic islands, the other for subsequent survival in a relatively impoverished but tightly selected bird community.

	São Tomé (Gulf of Guinea, Atlantic)	Anjouan (Comoros, Indian Ocean)
Herons	3	4
Cormorants	1	
Grebes		1
Rails	3	1
Ibises	1	
Eagles	1	3
Falcons		1
Owls	2	2
Quails	1	1
Kingfishers	1	1
Swifts	2	2
Pigeons	5	5
Cuckoos	1	
Rollers		1
Parrots	1	2
Shrikes	1	
Drongos		1
Flycatchers	1	1

	São Tomé (Gulf of Guinea, Atlantic)	Anjouan (Comoros, Indian Ocean)
Sunbirds	2	1
Orioles	1	
Bulbuls		1
Weaver-birds	6	3
Finches	1	
Starlings	1	
Crows		1
Warblers	2	
Thrushes	1	1
White-eyes	2	1

There is a strange beauty and logic in contemplating how an aerial voyaging dove got to be a heavy earthbound Dodo. The story is augmented and the logic reinforced by another extinct 'dodo', the Solitaire from a neighbouring island, Rodriguez, which developed rather different proportions. The two birds show how animals of the same stock can diverge on separate islands.

The climate, latitude and land-form of an island, its size and distance from land are generally what are immediately obvious and important to its human inhabitants or visitors. Also its character will be governed by the less accessible features which have guided the evolution of its animals and plants: geological age and origins, its soils and the direction of prevailing winds and currents. Above all the local animals and plants not only emphasize the peculiarity of each island; they also change it as an environment. The fact that every island differs in the mix of animals and plants that have colonized it helps ensure that most organisms will diverge from their cousins living in different communities and at different densities on other islands or continents. All these influences act on an older matrix and the fascination of dodos begins with their genealogical roots.

Some form of pigeon was among the first, perhaps the very first, to arrive on the Indian Ocean volcanoes that we now call the Mascarene Islands. This is unlikely to have happened much earlier than 100,000 years ago, which is when several million years of volcanic activity ceased on Mauritius: a relatively short period in evolutionary terms. Perhaps they were blown in from the east on trade winds, but it seems more probable that they originated in Madagascar, which is only 800 km from Mauritius. Whatever their origin, the transformation of these birds on the Mascarene Islands was so total that early naturalists wavered between classifying Dodos as Ostriches, vultures, waders, cranes or penguins.

The transformation is less extreme if we compare dodos not with adult doves but with young doves, or squabs. While squabs squat flightless in the security of their nest their abdomens are no less swollen and their tails and wings no more functional than those of the dodos. With neither predators to flee from nor competitors for food, there was less pressure on the Dodos' ancestors to grow *up* than to grow *big* so they became, in a sense, overgrown squabs.

But if one stranded pigeon becomes an infantile slob why should its neighbour become an athlete?

The colonist François Leguat, writing in 1693, left us a description of the Rodriguez Solitaire, *Pezophaps solitaria*,

so called because it is rarely seen in company and generally found alone, although it occurs in large numbers . . . the bird cannot use its wings to fly with, but employs them for fighting, for it has there a bone the size of a musket ball with which it strikes its adversary.

Before the island's discovery the only likely adversaries would have been other Solitaires, and the spacing out of pairs in territories was described by Leguat.

Corroboration for the Dodo's pigeon ancestry has come from some remote islands in the Pacific. On two Samoan islands there is a large dove, *Didunculus strigirostris*, which has become very terrestrial (although still able to fly). This dove's bill has become inflated to a degree that is intermediate between a conventional pigeon and the dodo. Yet again, on the island of São Tomé in the Gulf of Guinea there is a race of the African Green Pigeon, *Treron australis sao thomae*, where the horn of the beak is more expanded than in any mainland pigeon. (The tendency is widespread and seven other birds in the Gulf of Guinea have enlarged beaks; they range from finches to an oriole, a cuckoo to a starling.)

It is significant that most pigeons make wing-clapping displays and that Leguat also saw wing-clattering in Solitaires. Followed up by a real buffeting of the armoured wing bones, these very effective defence mechanisms may explain the relatively modest dimensions of the Solitaire's beak, which did not have to include defence and display among its functions. By contrast, Admiral Van Neck described the feeble-winged Mauritian dodos nesting on the ground in huge flocks in the 1590s. Interestingly, the early Dutch sailors observed that they were capable of ferocious pecks with their formidable beaks.

The longer legs that they possessed would have meant that solitaires were more active and mobile while relatively dispersed food on Rodriguez was implied by their solitary, well-spaced life-style. On the other hand, drastic reduction in the wings, shorter legs and denser packing of the Mauritian birds, together with their ability to lay down fat, suggests that there were very concentrated, abundant but probably rather seasonal, sources of food on Mauritius.

As the only large vegetarian animals the dodos' own numbers would have been the major limit on their food supply. Larger beaks may allow some species to eat more and faster, and so outstrip others of their own kind.

Dodos' beak sheaths were periodically shed, like antlers, so there may have been elements of display as well as domination or intimidation. There is no evidence that Dodos ate exceptionally tough foods so the huge disproportion of their beaks may illustrate, in an exaggerated form, a more general trait in many island birds. When the main competitors are members of the *same species* bigger beaks can contribute to survival.

Only on islands as remote and small as the Mascarenes could an early stranding of doves provide the genetic base for their development into large vegetarians. That the same stock could give rise to two such different creatures is a reminder that not only species but also the places they inhabit are truly unique.

Island evolution is thrown into bolder relief by such comparisons between two species. For Dodo and solitaire, being more or less social probably correlated with higher/lower densities, more/less food, greater/less stability of climate. Mauritius and Rodriguez differed in being older or newer, wetter or drier, differences that are the starting point for all island evolution.

Atlantic and Indian Oceans

The twelve groups of islands off Africa's shores have numerous such contrasts and although they cannot compare with the Galapagos or Indonesia as living showpieces of evolution they do not lack for interesting and significant species.

The first contrast, between east and west, Indian and Atlantic Oceans, coincides with numerous other differences, one being particularly significant. The Indian Ocean has been consistently warmer and more stable and its islands are older and larger land masses. It follows from this that their faunas and floras also tend to be older and more tropical. (However, the largest island, Madagascar, is a world of its own and is not our concern here.)

The islands of the western Indian Ocean.

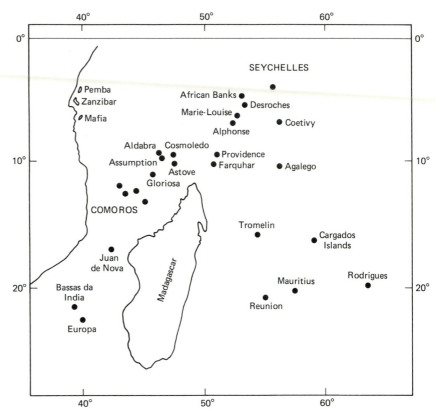

Socotra is the driest and most northerly of the Indian Ocean islands. The Seychelles are the oldest, Aldabra atoll is the newest, smallest and lowest. Zanzibar, Pemba and Mafia are close inshore, hot and wet. Of islands in the Comoros archipelago, some are nearer Africa, and others Madagascar.

The Atlantic, chilled by glaciers and polar currents, has had less stable temperatures than the Indian Ocean. Its islands, all volcanic, are of varying age. The southernmost, Tristan da Cunha, is the most distant from Africa. St Helena and Ascension are the most isolated of tropical Atlantic islands. The four truly equatorial Benin islands, graduated in sizes and distances from land, include the most mountainous, Bioko (Fernando Po), which rises to 3000 m and like St Helena has distinct bands of mountain vegetation on its slopes.

Saint Helena's daisy-trees. *From left,*
Melanodendron integrifolium,
Commidendron robustum, Petrobium
arboreum and *Psiadia rotundifolia.*

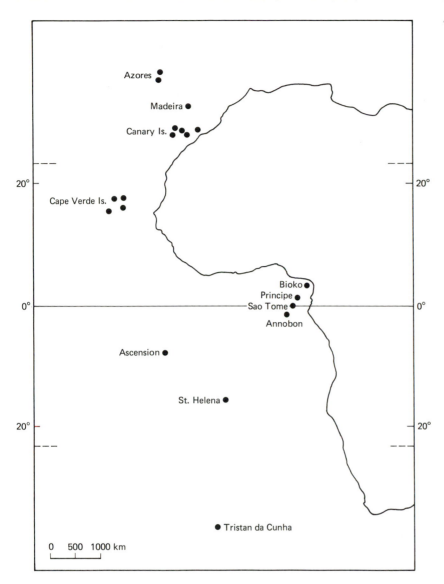

The Atlantic African islands.

St Helena is a single volcanic cone of 122 sq km that lies 1760 km from the coast of Angola. Once covered in tropical vegetation it is now two-thirds barren and eroded, with a population of over 5000 people, their stock and various feral animals. Among some one hundred endemic plants, several of which are now extinct, were forest trees. Three of these, *Commidendrum*, *Melanodendron* and *Petrobium*, are derived from the sunflower or daisy group, Asterae, which have very abundant, light and easily spread seeds.

The growth forms of these three genera, which have evolved on the island, resemble those of various unrelated mainland trees. The bastard gumwood, *Commidendrum*, evolved four species with open, layered canopies. *Petrobium arboreum*, at 7 m, is a tall tree for St Helena and its foliage forms a dense cone above a slender, straight bole. *Melanodendron integrifolium* is a flat-topped spreading tree with a dense network of branches springing out of a very thick, squat trunk. Here on a remote island, chance windfalls of minute seeds have led to a replay of the radiation of forest trees. Plants that would probably have remained herbs or shrubs on the mainland have given rise to the diverse architecture and enlarged body sizes that are typical of forest trees.

Amongst this singular flora and fauna St Helena also has the distinction of possessing the world's largest earwig, *Labidura herculeana*, an elongated flightless insect measuring 7.8 cm. It was last seen in 1965 but is now thought to be extinct. An expedition from London Zoo in 1988 found that a refuse dump close to the last known locality had introduced large numbers of mice, which had augmented their rubbish-tip diet with the earwigs. Only nibbled fragments of the insects could be found.

St Helena's beetle fauna numbers 256 species, 137 of which are endemic and most of which must have evolved from a very few beetles already on the island. How? The answer seems to be that insects readily develop a fixed association with a single food or host plant. Such plants often grow on special soil types, in an altitudinal belt or on only one face of the island, so that there might be a slight spatial separation. When the insect acquires a new host it is equivalent to the colonization of a new island. Thus three closely related species of *Longitarsus* beetles live near the summit of St Helena: *L. melissi*, on a type of cow parsley, *Sium*; *L. helenae* on *Lobelia scaevifolia*; and *Longitarsus janulus* on the groundsel *Senecio prenanthifolia*. This type of evolution is frequent in insects with exclusive hosts and even commoner for parasites where the host is still more of an island.

Invert the simile and an island can be seen to have some of the characteristics of a host organism. Its colonists soon adapt to the very particular pattern of its climate and geography but also find themselves among other animals and plants that constitute a community unlike any other, as specific an environment as that provided by a host species.

Viewing island communities as unique ecosystems, hosts to their component species, gives coherence to their study. The fact that physical isolation is normally essential for speciation and that every island differs, offering innumerable permutations for unique essays in adaptation, invests islands with their special interest and importance.

Islands seldom detain the great migrants such as storks, waterfowl, terns and others. The homing abilities of migrants seem to be too well developed to be subverted and perhaps the traditions and seasonality of their breeding are so well established that permanent settlement is inhibited.

There are, however, a few. For example, many rails and crakes make extensive migrations in spite of being poor fliers. Their journeys across continents and water are frequently punctuated and must be assisted by amphibious, semi-nocturnal habits and the use of loud calls to maintain contact.

Water rails are found all over the world and are known as fossils in the Paleocene, 62 million years ago. About the size of a domestic pigeon, the modern water rails of Africa and Eurasia have functional refinements for their stealthy existence in marshes, among them long toes on slim gymnastic legs and a slender beak. As with other continental animals, immense numbers of predators and competing animals have contributed to place rails tightly into a narrow ecological niche. One rail, *Aphanocrex podarces*, lived on St Helena until the island was settled in about 1510. Like the dodo on Mauritius, this rail was very large, and was the most important terrestrial animal. It probably subsisted mainly on the abundant invertebrates of this once-luxuriant volcano, perhaps also on land snails of which there were some 20 species.

On the rather barren island of Ascension the endemic rail was also flightless but of conventional size and the only indigenous resident land-bird there; it was last seen in 1656, long before the island was settled, so it seems it was probably exterminated by introduced rats. A giant rail also evolved on Mauritius, where they ran in numerous flocks. This species was last seen in 1693.

Modern *Rallus* water rails have developed from and replaced less highly specialized rails which preceded them. A surviving representative of this proto-water-rail lives today on Madagascar in the form of the White-throated Rail, *Dryolimnus cuvieri*. On Madagascar, where there are numerous predators, the birds can fly. On Mauritius the local race was flightless and, in common with many other birds, it became extinct about 200 years ago. The bird also succeeded in colonizing the Aldabran atolls and about 1000 of them still survive on Aldabra where they are effectively flightless, being only able to flutter up on to low rocks and branches. Aldabra's fauna and flora are not very old: the atolls have suffered periodic submersions, the last being some 50,000 years ago; this means that the White-throated Rail's almost complete loss of flight has taken place within this period of time.

If a bird can lose one of its fundamental characteristics in such a short space of evolutionary time there must be considerable advantages in doing so. What could they be?

Darwin considered the problem in relation to flightless beetles on Madeira and he suggested that flying forms could be more prone to being blown out to sea.

There is a simpler explanation. Flying is an energetic activity but an animal still burns up energy when it is not in flight, merely by maintaining unused flying apparatus which weighs a lot and takes up a great deal of space. Furthermore, the rest of the body, most particularly the legs and vertebral column, must be adapted to carrying about and balancing that burden of bone and muscle required for flight. What could be more awkwardly contrived than a pelican waddling on the ground?

A bird that relinquishes flight will actually *improve* its poise and efficiency as a terrestrial animal by reorganizing the architecture of its body and shifting its centre of gravity.

One of the most extreme examples of flightlessness in a bird is exemplified by the minute Tristan Rail, which still survives on a small islet in the south Atlantic. Appropriately named *Atlantisia rogersi*, it was discovered in 1923 by a Mr Rogers on Inaccessible Island, one of the Tristan da Cunha group, which lie 2700 km off Africa on the mid-Atlantic ridge.

Inaccessible is 12 sq km of cold wind-swept turf. What food there is lies under the dense canopy of grasses and miniature tree ferns. To be able to travel freely between the tussock bases, where there is shelter from the elements and potential predators, it is essential to be small. The climate and surroundings impose efficiency, while sparse food demands economy.

Economy can be achieved not only by using less material but also by not wasting what little there is, and this is one way of looking at the Tristan rail. In the course of miniaturization and redistributing its weight it has truly burnt its boats by dismantling the very mechanism that first brought its ancestors to Inaccessible. The sternum and pectoral muscles that power flight have all but disappeared. Thick plumage tends to disguise the bird's figure but cut-away profiles of *Atlantisia* and a flying rail (a Spotted Crake) reveal the extent to which it has adapted its stance and proportions.

Of the many migrant birds that must encounter the virgin slopes of oceanic volcanoes, rails seem to be the most able of terrestrial settlers. There is, however, an unusual instance of semi-terrestrial habits in a small bird, *Amaurocichla*, from São Tomé, one of the four Benin islands. Strung along a 600 km fault line running oblique to the Gabonese coastline, these islands erupted out of the sea some 20 million years ago and vary in size and ecology. At about 150 sq km Annobon is the smallest, the most recent and the most remote, being 340 km out to sea. São Tomé, a larger and more varied island of 1000 sq km, lies 280 km from Cape Lopez on the mainland and has the most peculiar fauna and flora of the group. It has been suggested that some stranded Palaearctic migrant belong-

São Tomé sunbirds. *Left*, the giant *Nectarinia thomensis, right* Newton's Sunbird *N. newtoni* (to same scale).

ing to the whitethroat or silvid group of warblers is the most probable ancestor of *Amaurocichla*. This long-billed, short-tailed warbler has become a very weak flier but has compensated by becoming agile and long-legged.

Amaurocichla is one of 17 bird species and 7 subspecies endemic to São Tomé. In a total of 40 species this is by far the highest level of avian endemism in any of the African oceanic islands and must relate as much to the age of the island as to its isolation. Another endemic bird, *Neospiza*, which may now be extinct, resembled one of Darwin's Galapagos finches in its development of a powerful parrot-like bill. A third endemic, the São Tomé weaver, *Ploceus grandis*, has been described as being like a nuthatch in the way it searches branches and tree trunks. Island finches diversify in the *size* of their beaks and food items, while mainland finches differentiate by habitat and the seed species they eat. Woodpeckers and nuthatches are weak fliers and appear to be poor colonists of islands, even one as close to land as Pemba or as large as Madagascar (where a nuthatch-like type has also developed from an unrelated endemic). São Tomé also has the distinction of harbouring the largest sunbird in Africa, *Nectarinia thomensis*. A much smaller sunbird, *N. newtoni*, has diverged less from its main-

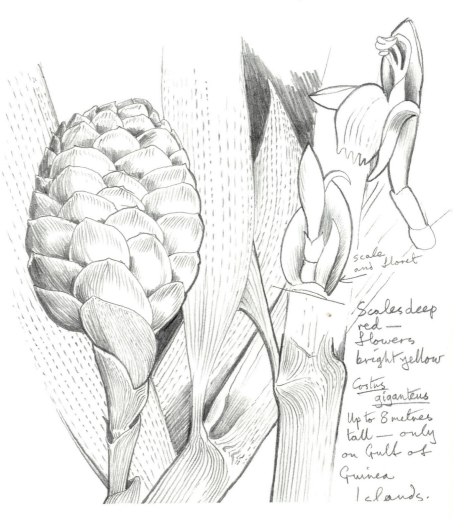

scale and floret

Scales deep red — flowers bright yellow

Costus giganteus

Up to 8 metres tall — only on Gulf of Guinea Islands.

Giant ginger *Costus giganteus*, from the Gulf of Guinea islands.

land parental stock and probably reached the island much later. Among the many interesting plants is a gigantic ginger, *Costus*, which has fronds up to 8 m in height.

An extraordinary colonization of this volcanic island concerns a caecilian, *Schistometopum thomensis*, which is about 30 cm long and often bright yellow in colour. Since salt water poisons amphibians the colonists must have rafted in enclosed in a log or some dense tangle of floating vegetation. Furthermore it is possible that Atlantic currents carried this raft in from the west because *Schistometopum* resembles the South American genus *Dermophis* more than any African form.

Two parrots, two starlings and two fruit-eating pigeons were once common on Príncipe, which is a smaller, less mountainous but moist island 146 km north of São Tomé. Devastated by dense and uncontrolled settlement, some of these woodland or forest fruit-eaters may no longer exist.

Like São Tomé, Príncipe has a warbler-like bird, *Horizorhinus*. Another enigmatic warbler is one of the 26 species of breeding land-birds on the Cape Verde Islands. Here too, 750 km beyond the most westerly tip of Africa's 'Green Cape', lives the world's most restricted bird species, the Raza Short-toed Lark, *Alauda razae*. About 50 pairs inhabit an area of less than 100 hectares on a rocky islet of 5 sq km. This would seem to be a type of social skylark in which the males grow massive beaks. Although they have been reported to dig in hard soils for the grass seed and grubs on which they feed, the enlarged beak seems to be more a reflection of male competition within the skylark flock.

Skylarks are no longer found south of Morocco but presumably they travelled much further south during the Ice Ages, at which time a population may have become stranded on the Cape Verde Islands. As temperatures rose, climate and vegetation would have become less and less favourable to temperate-adapted larks. In this tiny population social living would seem to have led to sexual selection in one feature, the beak. The birds were also progressively restricted to the only part of the island where they could out-compete three other desert larks.

Raza Lark, *Alauda razae*.

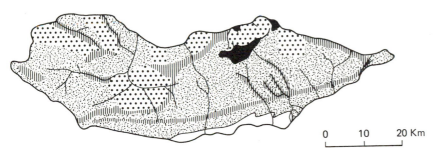

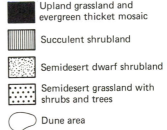

	Upland grassland and evergreen thicket mosaic
	Succulent shrubland
	Semidesert dwarf shrubland
	Semidesert grassland with shrubs and trees
	Dune area

Socotra Island.

On the other side of the continent, on the barren islands of Socotra and Abd el Kuri, the only endemic genus of bird, *Incana*, emphasizes how islands may conserve older types. Reg Moreau, a doyen of African ornithology, suggested that this bird might have remained on the island from the time it became detached from the Afro-Arabian continent, at least 10 million years ago. He thought it might represent a relict of the stock which gave rise to the grass warblers, *Cisticola*, now the most successful and diversified of all African warblers. Socotra has been settled for millennia and the animals and plants that remain represent a degraded fraction of what once existed. That rivers flowed there some 2000 years ago is reported in the Periplus of the Erythrean Sea (the account of Ptolemy's 2nd century AD voyage into the Arabian Sea). At that time

Socotra had no mammals but numerous crocodiles, a great many snakes and lizards of enormous size. Of this reptilian abundance thirteen genera remain; among the three endemics is a handsome large-headed gecko, but the island's greatest interest is botanical. Two hundred and sixteen plant species are endemic to the two islands, 85 of which are in immediate danger of extinction.

Three Socotra endemics, *Dendrosicyos*, *Dorstenia gigas* and *Dirachma*.

Amongst these are a sweet-smelling geranium-like shrub called *Dirachma*, the only member of its family, a verbena-like *Clerodendron leucophloeum* and the only representative of the cucumbers to grow in tree form, the weird, bulbous Cucumber Tree, *Dendrosicyos*. Two other plants have extraordinary growth forms. The first is the giant tree succulent, *Dorstenia*, where the inflorescence resembles an opened fig. In this plant minute male flowers are spread out over a broad-lobed receptacle in which the female flowers are embedded and when the seeds mature they are shot out of this receptacle over some distance. The second is the Abd el Kuri *Euphorbia*, which has leafless columnar stems which form colonies 'like a forest of green candles', all apparently linked by a single root-stock. Another endemic plant, the Socotran Pomegranate, *Punica protopunica*, was thought to be extinct but was found by Quentin Cronk in 1985 to be still growing in the Hamhil Mountains.

For centuries the island's major export was 'Socotra aloes', a drug extracted from the indigenous aloes and used as a purgative, as bitters and in cosmetics. In 325 BC Alexander the Great sent an expedition to Socotra to find out how Aloes was produced. Another source of ancient interest was 'dragon's blood', a brilliant vermilion resin extracted from *Dracaena cinnabari*, a Socotran endemic, and used as paint or protective varnish.

It is remarkable that many of these endemics have survived as long as they have. Isolated for at least 10 million years from any contact with large mammalian herbivores, it is clear that few of the endangered plants have evolved any ability to protect themselves from the onslaughts of camels, sheep and goats. Some of the survivors shelter beneath tangles that are distasteful to livestock, like the endemic vine, *Cissus subaphylla*. Over the last few centuries livestock too have found it more and more difficult to survive. Where once there were rivers there are now only sand gullies. In the 13th century there was no shortage of milk and meat. In 1612 the island was still sufficiently moist to support water buffaloes—now it is camels. Many of the last plants hang on among boulders on a few high, steep slopes which act as moisture traps.

In spite of being 250 km from the mainland, 2000 years of settlement has given people ample time to strip Socotra, degrading or obliterating the fruits of 10 million years of isolation and self-contained evolution.

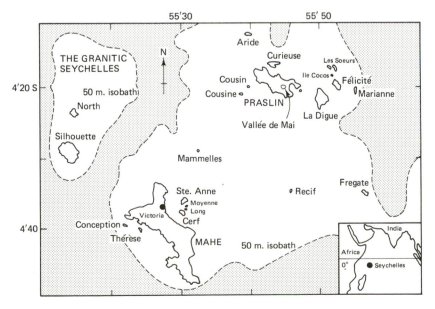

The granitic Seychelles with 50m isobath to indicate the former extent of land.

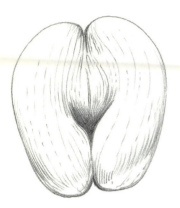

Coco-de-mer coconut, *Lodoicea maldivica*.

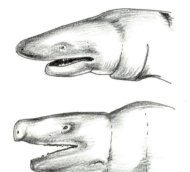

Two Caecilians from Seychelles, *above*, *Hypogeophis rostratus*, *below*, *Grandisonia seychellensis*.

Seychelles frog, *Sooglossus seychellensis*. Male with tadpoles attached to its back.

Socotra's period of detachment from Africa and Arabia has been brief compared to that of the Seychelles. Nearly 3000 km from India, 1300 km from the African coast and isolated for 75 million years, these islands have over 70 endemic land animals. Today's islands are the hilltops of a much larger land mass, 55,000 sq km in extent, that was above the surface of the sea during the Ice Ages, 16,000 years ago, when sea levels were 130 m lower than at present. When General Gordon visited the islands in 1881 they had been settled for little more than 100 years and he could still say that here was the Garden of Eden, complete with fruits of a gigantic tree of knowledge in the form of sexually suggestive coconuts.

Although most species are at risk in the Seychelles, a relatively short period of settlement has led to the extermination of fewer species than on most oceanic islands. None the less, people have destroyed most of the large trees, all the giant land tortoises and crocodiles that were so abundant 200 years ago, as well as exterminating a parrot and a white-eye.

The clearest sign of the great age of the Seychelles comes from those very primitive trees, palms of which there are six endemics, the best known of which is the coco de mer, *Lodoicea maldivica*, a very tall slender species with straight boles going up to 30 m. The leaves can be up to 6 m long and 4 m wide. They grow very slowly, as do the fruits, which at 22 kg are the heaviest seeds known. Mature trees are perhaps 1000 years old.

Another plant only found in the Seychelles is the Jellyfish Tree, *Medusagyne*, so called because the shape of the seed clusters is reminiscent of a jellyfish. This is a small tree with leathery leaves and diminutive white flowers. Only about 11 plants are known to exist in the wild, growing in deep clefts high up between granite domes. Rooted in deep clefts, they are unusual in being able to resist drought and in being dispersed by wind (most island plants are bird or bat dispersed). This combination is scarcely conducive to survival on a small moist island and suggests that *Medusagyne*'s original adaptation was to a larger and drier land mass. An unsubmerged Seychelles bank during drier periods, or dry parts of Gondwanaland perhaps? Its affinities are unknown and *Medusagyne* is likely to be an exceptionally ancient plant. The palm *Phoenicophorium* is another Seychelles endemic that can colonize dry eroded surfaces, but by far the larger part of the island's plants are moisture-loving.

Another rare endemic is a wild vanilla orchid which grows in twining leafless masses and bears pretty pink-lined white flowers. Of the animals, six species of caecilian (out of a world total of 100) are perhaps the most significant. These primitive legless burrowing amphibians are likely to be 'Gondwana animals' left over from Seychelles' period of attachment to Africa (it is, however, remotely possible that their ancestors could have rafted to the islands). They spend the embryonic stage as gilled tadpoles within the egg; these are laid in moist soil and guarded by the female coiling her snake-like form around the gelatinous cache. Other amphibians which may also be Gondwanan are the Sooglossid frogs typified by *Sooglossus seychellensis*, where the tadpoles mature attached to the back of male adults. There is also a single endemic species of chameleon which must have rafted to the islands from Africa or Madagascar in quite recent times.

The absence of large herbivores allowed tortoises (probably an early form of African or Indian land tortoise) to become the major grazers on at least twenty western Indian Ocean islands, where early mariners exploited them for food. In 1776 the French explorer Marion de Fresne took five Seychelles tortoises, *Dipsochelus arnoldi* (a species that subsequently became extinct), back to his base in Mauritius, where what was thought to be the last of them died in the military barracks in 1918. This individual could have been in the region of 200 years old. Since then two living males have been found in the possession of Mauritians, which may also derive from the original de Fresne group. Giant tortoises of a closely related species *Dipsochelus elephantina*, still survive in great numbers on Aldabra, an atoll which was probably colonized by animals floating in from the Seychelles. What distinguishes these oceanic giants from the mainland African tortoise, *Testudo*, is an enlargement of the nasal chamber and development of a valve that enables them to suck in water or drink through the nostrils.

The local flying fox, *Pteropus seychellensis*, has allied but distant populations on Aldabra, the Comoros Islands and on Mafia (a low island just offshore from the Rufigi River delta in Tanzania). Another species, the Pemba Flying Fox, *P. voeltskowi*, is exclusive to Pemba, where it is within sight of the mainland yet never visits it.

Giant tortoise from Aldabra.

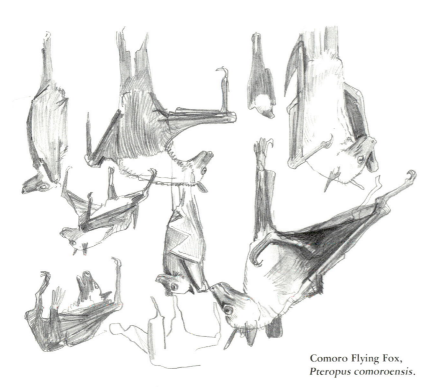

Comoro Flying Fox,
Pteropus comoroensis.

The very numerous races and species of flying fox are distributed through all the islands of the Indian Ocean as well as much of the tropical Pacific. The centre of their overall range is Indonesia and its thousands of islands, which is where they presumably first evolved as a distinct form of fruitbat. By the time populations had reached the east African coastal islands they had proved themselves great oceanic travellers, but were also clearly island specialists unable to compete with continental bats. Dependence on sea water for dietary salts could be one factor limiting them to sea-shores.

The good dispersers, like pigeons and rails, are early colonists of islands, but once there, island dwelling soon selects for stay-at-homes. This is partly because the great oceans are seldom kind to wanderers, but there are other changes to be made. On islands, bats must live within their means and balance their numbers with a crop of fruit that is generally made more reliable and predictable by an equable rainfall.

Social and reproductive behaviour is likely to become subtly transformed as the bats follow a unique island routine and breed in large roosts. On each island group bats will evolve effective strategies to exploit a finite supply of food with a very specific pattern of fruiting and distribution.

The islands of Pemba and Zanzibar are superficially similar but have one vital difference. Pemba is separated from the mainland by a trench 800 m deep which is at least several million years old. Zanzibar instead is likely to have had connections as recently as 50,000 and again 10,000 years ago across its shallow channel. Zanzibar has two rare and peculiar animals, which are not on Pemba or on the mainland opposite, but do recur in a narrow coastal enclave further north. One of these is *Cephalophus adersi*, a dwarf red duiker, the other is a very beautiful turaco, *T. fischeri*.

Perhaps the best known of Zanzibar endemics is its very ornate Leopard, which is sprinkled with an abundance of minuscule rosettes. The Zanzibar animals pack on two or three times as many rosettes as a conventional Leopard. This could be a 'primitive' type of pattern, but there is nothing else in the anatomy of the animal or the history of the island to suggest very great age and it seems more likely that the 'founder' of this population had an aberrant pattern type which has become fixed in what must be a tiny genetic pool, now numbering a handful of individuals.

Pemba is smaller than Zanzibar and more impoverished, for instance it has only 73 breeding land birds to Zanzibar's 105, but it has several distinct birds including a russet form of Scop's Owlet, a green pigeon and a very common white-eye, *Zosterops vaughani*, which differs substantially from birds on the mainland opposite. White-eyes, like pigeons, rails and ibises, recur prominently in lists of endemics, whether on islands or on mountains in mainland Africa. They are classic new or neoendemics. All are successful colonists because they belong to versatile wide-ranging types of birds but unlike conventional migrants they are also able to settle down as residents in localities that suit them. White-eyes in particular seem to be winners when it comes to colonizing wooded habitats. Extremely thorough foraging for minuscule foods will always extract sustenance for such tiny and active birds, no matter how rich or degraded the vegetation. Perhaps even more important, they quickly adjust to the biological peculiarities of a place and its climate. Depending on the nature of the competition, white-eyes can expand into a broader niche or squeeze into a narrower one. They share with sunbirds the speciality of being very active nectar and insect feeders but they also go for many other high-energy foods, such as the smallest seeds and soft fruits. Their tongues have a brush for lapping nectar and juice but the beak has not lost its versatility and their behaviour too can range from detailed combing of foliage to flitting from flower to flower.

The family's centre of distribution lies in Indonesia where most of the nearly 80 species occur. They resemble flying foxes in being exceptionally effective colonizers of islands where they then settle and speciate. Unlike flying foxes, white-eyes occur on the continental mainland but there is generally only one species in any single habitat. Although those at higher altitudes tend to be larger and have relatively large wings (possibly in response to reduced air density) and birds from drier areas are paler, greyer or browner than the darker, greener or yellower birds from moist areas, mainland white-eyes keep within narrow limits of variation.

Island birds are less constrained in their evolution. For example, the tiny Benin island of Annobon has only two passerines, one of which is a white-eye. This bird has become decidedly bigger than any mainland bird. Similar enlargement has taken place in all the other Benin island white-eyes, but evolution has proceeded so far that the birds are classified as another genus, *Spierops*. (One of these, *S. melanocephala*, has actually colonized the mainland, living between 1800 and 2700 m in the uplands of Mt Cameroon.)

The upland habitat on Grand Comoro is very restricted but here too there is a

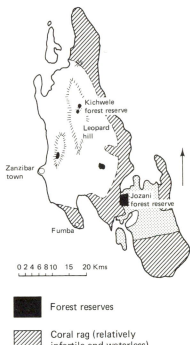

Zanzibar Island.

Dwarf Red Duiker *Cephalophus adersi* from Zanzibar.

The sub-species of *Charaxes candiope*. *From top to second from bottom, C.c. thomassius* (São Tomé), *C.c. velox* (Socotra), *C.c. candiope* (Continental Africa). *Bottom Charaxes analava, an allied species from Madagascar.*

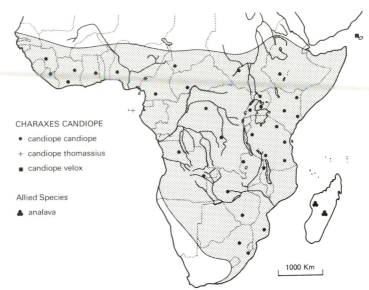

CHARAXES CANDIOPE
• candiope candiope
+ candiope thomassius
■ candiope velox

Allied Species
♣ analava

1000 Km

The distribution of three sub-species of *Charaxes candiope*.

large green white-eye, *Z. mouronensis*, which is specific to vegetation above 1500 m, while another smaller species is common below. The drier north-east African mountains are exactly equivalent in sheltering a large bird, while lowlands support a smaller species.

In all these cases it would seem that the first colonist has become larger only to be eventually displaced from its original niche by a later invader that has pushed it into the more difficult or marginal extremities of its range. This is a pattern that is repeated for many other organisms on the mainland and it helps to explain why archaic forms tend to be stranded on mountains while their advanced relations are more widespread.

One of the best examples of genetic change in isolated populations comes from the *Charaxes* butterflies. On the continent the green-veined *Charaxes candiope* ranges over most wooded habitats. Although somewhat variable in pattern it is considered to be a single subspecies. This butterfly has colonized São Tomé, Socotra and Madagascar. On the last island it has speciated into five forms, the most distinctive being *C. analava*, while the São Tomé and Socotra species have also altered their underwing patterns. These butterflies rest and feed with the wings closed, and the underside resembles a curled and mouldy leaf in the mainland insect. Alterations in pattern result from contours or contrasts migrating back and forth in between the veins or ribs of the wing (as though they were soap films caught in a spider's web) and by local alterations in tone and colour intensity. In the smaller Socotra insect, *C. c. velox*, these elements are more aligned and form regular 'cells' that probably match dry stems and twigs no less effectively than they do leaves. On São Tomé, *C. c. tomassius*'s tones simplify into a bold contrast of black and orange that accords with Gloger's rule (which observes that tones darken in humid environments). The Madagascan *C. analava* pattern is not so much a first-order imitation of a dead leaf as a second-order mimicry that combines the protection of a larger-scale cryptic pattern organized into bold snake-like curves and patches, with the further protection of a reptilian eye-spot. The patterns and strategy resemble those of the emperor moths, Saturnidae. Should the camouflage be penetrated, the potential predator's last-moment perception of a pseudo-snake eye may well prevent predation. This could very well help to protect the insect from birds and lemurs that may also converge on the butterflies' feeding sites (resin scars, fermenting fruit-falls and salt deposits). In any event, the different wing patterns imply different selection pressure on islands, where founders are few in number and gene-pools smaller, than for a continental population. The fact that *C. candiope* is one of several dozen *Charaxes* species on the continent, yet one of very few butterflies on an island, illustrates the fact that continental populations tend to occupy much narrower niches, even if the habitats and climates in which they live are more diverse than on an island.

It is interesting that the island of Pemba also boasts a local *Charaxes* butterfly, but from another mainland group.

Concern and Conservation

Public interest in island fauna and flora existed in Europe from the time of the early explorers. On the other hand, concern for the survival of island species began much later. Darwin and Wallace first proclaimed the evolutionary significance of islands, but, notwithstanding Darwin's plea for the preservation of Aldabra, the intellectual interest of their work was seldom matched by concern for the survival of island communities. Nonetheless attempts to check uncontrolled destruction of island flora and fauna did begin more than 200 years ago. For example, the first nature protection laws for Mauritius were made in 1767 and the rarer Mauritian birds were put under strict legal protection in 1878. A Mauritius Natural History Society was founded in 1829 (and still exists). An ambitious conservation action plan was drawn up within Mauritius in 1950. It is very important to remember that such expressions of concern have a long history. Yet it must also be admitted that this has scarcely slowed the pace of destruction. Why?

Such initiatives can always be traced to eccentric individuals. However, in the face of general indifference, sometimes outright hostility, the influence of such naturalists has seldom spread outside their immediate circle. Although individuals have always been and always will be crucial in drawing attention to the plight of indigenous island communities it is only when the franchise for conservation enlarges, when conservation becomes a part of popular culture, that there is any prospect of checking the decline.

There are at last signs that this is happening. In the 1960s Gerald Durrell founded the Jersey Wildlife Preservation Trust with the dodo as its symbol. The Trust now has an international membership of 15,000 and supports a programme of financial aid, education, training, research and captive breeding for the benefit of endangered Mauritian wildlife. Such intense international interest so heartened conservationists within Mauritius that they were able to set up their first major nature reserve in 1974. Since then the forestry department, in close association with the WWF, IUCN and ICBP, have mounted a very ambitious plan to conserve the 400 endemic plants of the Mascarenes.

Interest in island birds has been fuelled by a long tradition of island expeditions by the British Ornithological Union (including one to Mauritius). The navy base on Ascension provided facilities for an expedition led by Stonehouse in 1957. C. Benson and another team studied the Comoros' birds in 1958. Each of these expeditions greatly increased knowledge of the birds but also aroused efforts to promote conservation in the islands. An important vehicle for this was a small society that was founded early this century to lobby for parks, game reserves and protection laws in what was then the British Empire. Now known as the Fauna and Flora Preservation Society, it still lobbies for conservation and supports projects, particularly in English-speaking countries and especially endangered island communities. Both FFPS and WWF were involved in an intensive lobbying, press and TV campaign during the 1960s to dissuade the British Government from leasing Aldabra to the USA as a strategic air base. The

Pemba Charaxes, *Charaxes pembae*.

campaign was successful and the British scientific body, the Royal Society, was charged with designing and implementing detailed field studies on the vegetation and animals of Aldabra, notably the giant tortoises. Following the ceding of Aldabra to the independent Government of Seychelles this scientific study and the protected status of the islands have continued. The Seychelles Government is exceptionally conservation conscious. It has declared Praslin's Vallée de Mai (the principal locality for Coco de Mer) a World Heritage Site and made the Peak of Morne Seychellois into a National Park. (The government has also been active in trying to exert pressure on the whaling nations.) The discreet influence of FFPS has been wide-ranging and in 1981 it was instrumental in setting up an important programme for St Helena. This scheme could be a model not only for small oceanic islands but also for other concentrated centres of endemism. It was no accident that FFPS initiated this programme as it came into being to persuade authorities and bureaucracies that had no expertise or tradition of responsibility for the environment and conservation. St Helena and Ascension have been British dependencies for hundreds of years, yet until 1981 no serious effort had been made to conserve the islands' unique flora and fauna.

Once covered in dense, semi-tropical forest, two-thirds of St Helena has become barren; the remaining 40 sq km is occupied by some 5500 people, their crops and livestock. The neglect of St Helena originally reflected the rudimentary character of the island's administrative structures and the absence of any expressed concern or pressure from people who might be aware of the island's biological importance. Following a preliminary survey by N. Kerr and R. O. W. Williams in 1970 the English botanist Quentin Cronk visited the island in 1980, when he not only confirmed that some 18 endemic species were on the point of extinction but also 'rediscovered' species that were long thought to be completely extinct, notably the St Helena Ebony, *Trochetiopsis melanoxylon*, and three species of gumwood, *Commidendrum rotundifolium*, *C. spurium* and *C. burchelli*.

On returning to Britain Cronk enlisted Gren Lucas and Clive Jermy of Kew, both on the Council of FFPS, to get backing for a plan of action. Through their initiative, funds were raised from FFPS, other conservation societies and the British aid body, Overseas Development Agency. In 1981 Cronk returned to St Helena and large numbers of seeds and cuttings of all the endangered species were propagated. A major objective of the action plan was to identify suitable conservation reserves where the reintroduction of indigenous vegetation had a reasonable chance of success. In areas where reintroduced vegetation took root, soil erosion could be checked and the barrens gradually regrown with indigenous vegetation using hardy St Helena scrubwood as the colonizer and as a basis for the plant succession that should follow. The project does not provide food or any immediate income but it should eventually help to bring back moist soils and something of the island's ecological diversity. Above all it holds out the hope that people may once again be able to see and study 'the St Helena microcosm'. The newly built St Helena Central School is being landscaped with endemic plants, the agriculture and forestry departments now have the propagation of endemic plants as official policy and organize nature rambles, public meetings, projects and expeditions to reserves. Schools now teach ecology with field studies in the biology of indigenous flora and fauna. Kew offers school leavers courses in horticulture and the first St Helenan to take a Kew course, G. Benjamin, is now back on the island with new skills and perceptions to pass on. St Helenans now know their home is truly unique.

The individuals who have set this revival in motion deserve great credit but their deepest satisfaction must lie in having brought renewal to a small but singular corner of the planet. Now what of the many other islands? Who will open the eyes of islanders in São Tomé and Bioko to the treasures they possess? Who will help the Yemenis halt the drying out of Socotra?

CHAPTER 2

The Cape of Good Hope
and Maputo Coast

The sea's pervasive influence on terrestrial communities is nowhere more obvious than at the southern end of the continent. Mountains run eastward from the Cape of Good Hope until they are cut by the wide valley of the Limpopo. The heights can be cold, their south-eastern scarps often precipitous and the coastal lowlands are nowhere more than 100 km wide (and then only to the north where the Maputo River, draining Zululand, spills into the Limpopo delta). It is the equatorial Mozambique current, running closely inshore, that warms and moistens this narrow littoral, but its influence ends abruptly at the Cape where it is turned by the icy Antarctic waters of the powerful Benguela current.

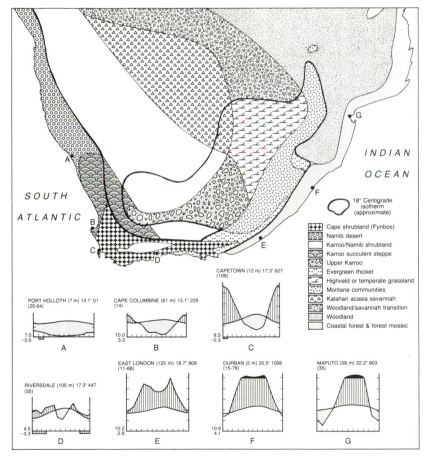

The influence of five major factors on natural habitats in Southern Africa. A Latitude. Cape shrubland or Fynbos in winter rainfall area (also Highveld in temperate uplands). B Altitude. Highveld on upland plateau. Montane communities on higher more easterly mountins. C Atlantic aridity. Longitudinally graduated belts of Namib desert along coast, Karroo/Namib shrubland and Kalahari acacia savanna inland. Latitudinally graduated from the arid western Karroo to the south-eastern Upper Karroo. D Indian Ocean Humidity. Longitudinally graduated belts of lowland forest (or forest mosaic) through woodland to Kalahari acacia savanna. Altitudinally graduated from lowland forest through evergreen thicket to montane communities (including forest). E Temperature. Operates within all above influences. The 18° Centigrade isotherm (for approximate mean annual surface temperature) is indicated in heavy black line.

Table mountain, Cape of Good Hope.

Cape Aloe, *Aloe ferox*.

The Namibian deserts are discussed later but the southern tip of Africa, the Cape, grows a unique flora that is very largely the product of a complex interaction between dry cold from the west, moist warmth from the east and winter rain and snow in the extreme south. Global changes in the past have tipped the balance in favour of one or the other of the three offshore climates and brought more or less cold but never to the point of producing glaciers. The boundaries of vegetation zones are known to have made small shifts along the coastlines and up and down slopes inland. Even so we can be sure that the two shorelines have each been brushed by their respective equatorial or polar currents for more than 60 million years. While contrasting climates have been a challenge to plants and animals over all this period of time, the two extremes have also met along countless narrow margins up and down the crags, ridges and valleys of this southern peninsula.

Between cold desert and tropical forest lie numerous intermediate states. Elsewhere these are major vegetation belts: here they have been compressed into a coastal strip of scarps, valleys and shorelines, some steep and barren, others flat and fertile. Like the Namibian coast this seaboard is some 2000 km long. Because this three-climate interaction is both very old and, within tight limits, very stable, the Cape and Maputo coasts have become sites for botanical radiations more concentrated than anywhere on earth.

The Cape itself is subject to two seasonal contrasts that are unique in Africa. Summer droughts are reinforced by very strong desiccating winds which turn the landscape into a tinderbox that only needs a spark from lightning to set it alight. By contrast, low winter temperatures delay evaporation so that plants can make the very best use of relatively meagre rain. Consequently the Cape flora has nearly ideal conditions for growth in winter, but must adapt to fierce summer fires. The product is 'fynbos', low bush dominated by heaths, proteas, sedges and rushes (Restionaceae) and many bulbous flowers such as irises and lilies.

Being exceptionally well adapted to the local climate and soils explains the pre-eminence of these families. Their further explosion into hundreds of species has been promoted by the very peculiar structure of a terrain that is dissected up into a mosaic of 'cells' within a region of three climatic extremes. The mountains and valleys of the Cape have served as 'population traps', each with its own enclosed environment yet each positioned within a larger gradient. Each minor shift in the boundaries has allowed some species to escape their particular redoubt while others have remained firmly parochial.

The Cape flora and fauna have always been isolated from tropical Africa by a succession of barriers. Only the most adaptable and generalized of animals and plants have overcome all these hurdles. Those limited by aridity have had the full breadth of the Kalahari, Karroo and Namib to contend with. Those limited by cold or by barren soils were repelled by the South African uplands. Those unable to cross rough terrain faced mountains. Broad perennial rivers, the Zambezi and Limpopo, inhibited others. These landward barriers were also inhibiting or delaying invasion from the north. Emigration was made even less likely because the climatic conditions of the south are reproduced, but only very approximately, in limited and distant areas of upland in eastern Africa.

Many older textbooks and atlases show the Cape with a 'Mediterranean' climate and environment; like the inappropriate naming of African animals after European models, this is primarily a reflection of the origin of early biologists and geographers. It is true that the climate suits vines and other Mediterranean plants. The grey bush that covers the slopes above Paarl resembles the 'macchia'

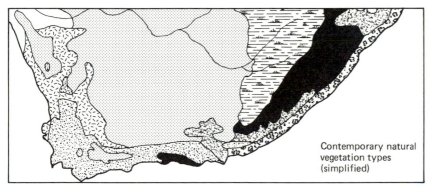

Contemporary natural vegetation types (simplified)

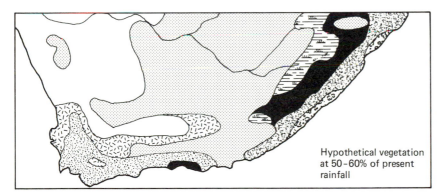

Hypothetical vegetation at 50-60% of present rainfall

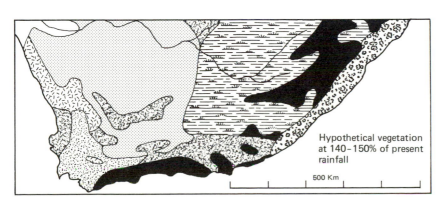

Hypothetical vegetation at 140-150% of present rainfall

500 Km

The migration of Cape vegetation zones with climatic change (modified after Cooke, 1964). *Above*, contemporary natural vegetation types. *Middle*, hypothetical vegetation at 50–60% of present rainfall. *Below*, hypothetical vegetation at 140–150% of present rainfall. The local patterns of fragmentation both in the past and present would have been much more complex than can be shown in these simplified maps. Within 100 km of the coast all mountain ranges would have carried a mosaic of vegetation that became most highly dissected in the south western Cape.

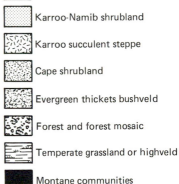

Namib desert

Karroo-Namib shrubland

Karroo succulent steppe

Cape shrubland

Evergreen thickets bushveld

Forest and forest mosaic

Temperate grassland or highveld

Montane communities

of Provence in being fire-induced, hard-leaved species and heaths flourish in both regions, but it is a misleading resemblance. For one thing, the Mediterranean is much poorer in number of species and, of course, there are no proteas and none of the tight mosaic of different communities that interrupt each valley and cliff face of the Cape. Sadly the Mediterranean simile has become more and more justified as vineyards, farms and villages elbow the indigenous vegetation into ever smaller enclaves.

The other great difference is stability. Global fluctuations in climate meant mass extinction in Europe during the Ice Ages. In southern Africa moving a few kilometres to the east or west was all that was needed to survive. Now that the unique climatic stability of the Cape and its lack of disturbance over some 60 million years has been recognized, the region has come to be identified as a major focal point for the evolution of flora. There are known to be 7000 species of plants in an area the size of Wales, more than half of which are endemic. Within this vast profusion of plants it is important to recognize that the isolation of this continental cul-de-sac has not only allowed an ancient residual fauna and flora to survive, some with little change, but peculiar climates and soils have also challenged more advanced plants, some derived from northern temperate regions, to generate locally adapted forms. The former can be called old or 'relict endemics', the specialists 'new endemics'. Typical of old endemics are the cycads represented in South Africa by the small fern-like *Stangeria* and the tree-cycads *Encephalartos*. Cycads are gymnosperms, the most primitive of seed-producing plants and a dominant form of life between 300 and 200 million years ago. Cycads grow over all the eastern margins of the southern African uplands. Most *Encephalartos* are rather similar but 26 forms have been recognized. This variety is all the more impressive when the longevity of these plants is taken into account. One specimen of the Natal cycad, *Encephalartos natalensis*, growing near the Valley of a Thousand Hills, has a root-stock estimated to be 1000 years old. Significant genetic change in such long-lived plants is likely to need millions, not thousands of years to appear. Although cycads are very poor dispersers and have been in steady decline for more than a hundred million years they are evidently quite resilient in the few stable areas in which they survive. The nearest relatives of this type of cycad only occur in Mexico and Australia.

Cycads are tough, and resistant to cold, drought and fire, characteristics that are also typical of other southern endemics, notably the heaths and proteas.

The proliferation of plant life can be illustrated with a selection of 20 of the 69 *Protea* that are endemic to South Africa. The distribution of these localized species are strung out east to west in successions that reflect gradients or local differences in temperature, humidity, soils, drainage and the influence of fire (see illustration and caption). As with other organisms, it is likely that climatic shifts isolated populations and caused the multiplication of species we see today. Proteas are peculiarly well adapted to local conditions, because they are insulated by their bark, and their roots can take up moisture on cool, well-drained and infertile slopes. To some extent they control conditions because a mixture of many species will grow in dense, low thickets which shade the soil and reduce evaporation, temperature and run-offs as well as deterring large animals which would compress the well-drained acid soils. Proteas derive from a rather primitive family with relatives in Australia and South America, but all these southern hemisphere relicts have continued to speciate and so cannot be easily called old or new endemics. The latter are better examplified by another group of plants that has adapted to survive long hot summers and cold winters, conditions more typical of the western Cape than the east. *Moraea* are flowers that probably derive from a tropical woodland iris, similar to the living *Dietes*. They have a corm that allows long periods of quiescence within a fibrous tunic that protects this storage organ from being dried out or attacked by insects or disease. Both

Head and mouth of *Peripatopsis capensis*.

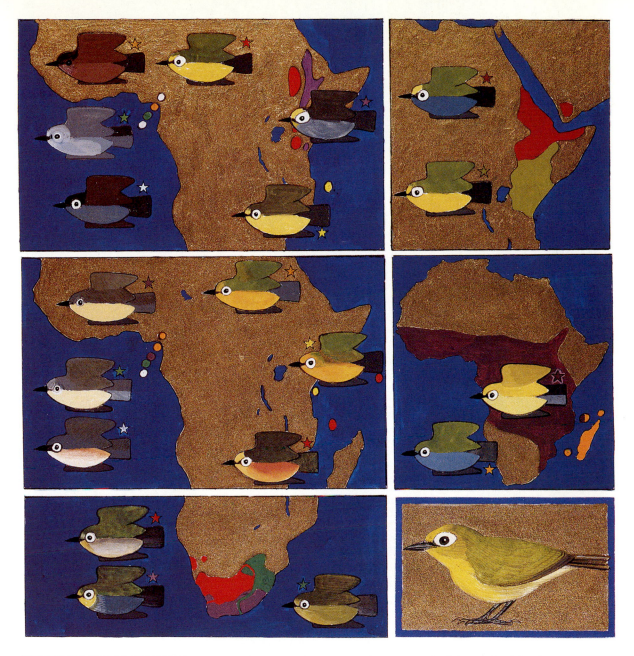

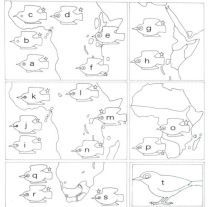

White-eyes (*Zosterops* and *Speirops*) of Africa and its islands

The colour of the star adjacent to the bird represents the colour of the distribution for each species

More differentiated and montane specialist white-eyes

a) *S. lugubris* (São Tomé and Mt Cameroon) **b)** *S. leucophea* (Príncipe)
c) *S. brunnes* (Fernando Po) *Z. poliogastra*
d) yellow belly form (NE Highlands)
e) grey belly form (NE Highlands)
f) Mauron Mountain White-eye *Z. mauronensis* (Grand Comoro)

One of the Horn of Africa lowland isolates, Z. abyssinica

g) Northern type, *Z. a. abyssinica*
h) Southern type, *Z. a. jubaensis*

Younger, less differentiated island White-eyes

i) *Z. griseovirescens* (Annobon) **j)** *Z. feae*

(São Tomé) **k)** *Z. ficedulina* (Príncipe)
l) *Z. stenocrita* (Fernando Po and Mount Cameroon) **m)** Pemba Island White-eye **n)** Mayotte Island White-eye *Z. mayottensis* (Mayotte and Seychelles)

The meeting of two species, both present on the Grand Comoro Islands

o) Yellow White-eye *Z. senegalensis*, the dominant species in Africa
p) Madagascar White-eye *Z. madraspatensis*

Cape Isolates

q) Pallid White-eye **r)** Cape White-eye **s)** Green White-eye
t) Pemba White-eye

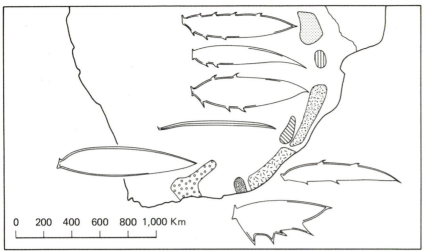

Legend for cycad map:

- *Encephalartos transvenosus,* Modjadi cycad
- *Encephalartos paucidentatus,* Barberton cycad
- *Encephalartos natalensis,* Natal cycad
- *Encephalartos ghellinckii,* Drakensberg cycad
- *Encephalartos altenstenii,* Eastern Cape cycad
- *Encephalartos latifrons,* Albany cycad
- *Encephalartos longifolius,* Suurberg cycad

Some endemic cycads, *Encephadartos* species, from the Cape and Maputo coast. The Modjadi cycad, Barberton, Natal, Drakensberg, Eastern Cape, Albany and Suurberg cycads.

the flower and the single leaf emerge at irregular, often long, intervals of time and a major trend has been from rather ephemeral flowers towards longer-lasting and very gaudy ones. Flowering probably correlates closely with the activity of specific insects, especially bees. The colour plate shows a small selection of the rarer relict species of the western Cape. Unlike proteas and cycads with their east–west gradients, the divisions that have led to the speciation of *Moraea* have centred more on an aridity gradient that runs north–south. Temperature, soils and altitude must also have played a part.

An indisputable archaeoendemic and relict family, the Achariaceae, is unique to the Maputo Coast and has only three species but is of particular interest because it is thought to provide an evolutionary link between the passion flower and gourd families. Gladioli, freesia, red-hot pokers (*Kniphophia*), crane flowers, *Strelitzia*, geraniums (*Pelargonium*), agapanthus lilies and the amaryllis, *Nerine*, are all flowers that originate in the Cape. They are now commoner in gardens than

Opposite page. Sugarbush flowers, *Protea* spp. Eighteen species with limited distributions. (Note how distributions narrow progressively from east to south-west). a *Protea inopina,* b *P. cryophila,* c *P. revoluta,* d *P. holocerica,* e *P. canaliculata,* f *P. stokoei,* g *P. angustata,* h *P. pudens,* i *P. aristata,* j *P. lanceolata,* k *P. intonsa,* l *P. mundii,* m *P. foliosa,* n *P. simplex,* o *P. dracomontana,* p *P. nubigena,* q *P. parvula,* r *P. comptoni.*

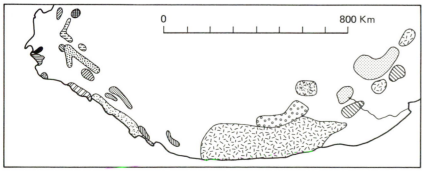

Legend for protea map:

inopina	pudeas
stokoci	foliosa
intonsa	parvula
dracomontana	holocerica
cryophila	aristata
angustata	simplex
mundii	comptoni
nubigena	canaliculata
revolta	lanceolata

Silver ghost moths, *Leto venus*, mating.

Cape moths. *Top, Spiramiopsis comma, middle, Brephos festiva, lower, Estigmene internigralis.*

The stag beetles, *Colophon* (Southern stag beetles have elaborated the forelegs rather than the head for ritualised fighting). 1. *Colophon westwoodi*, with right foreleg (anterior tibia) in black. Forelegs of 2. *C. stokoei*, 3. *C. cameroni*, 4. *C. thunbergi*, 5. *C. izardi*, 6. *C. primosi*, 7. *C. neli*, 8. *C, berrisfordi*. The numbers on the map indicate the respective species's isolated distribution on mountain massifs in Cape.

in their natural habitat, because most of their original range has been taken over by agriculture.

Among the most ancient but also the least prepossessing of Cape endemics are the twelve species of *Peripatopsis*. These are soft-bodied, velvet-textured invertebrates that look like khaki caterpillars with 18 pairs of clawed legs. About 5 cm long, they have antennae above a jawed mouth which is supplied with slime glands. They cannot close their spiracles and so desiccate in a matter of hours. They live in cool damp soil under logs, stones and in caves in the coastal forests. Their special interest for biologists is that they link two of the basic phyla of animals, the Arthropods and the segmented worms, and that similar invertebrates are known in fossil form from the Cambrian era more than 500 million years ago. Their survival in southern Africa is some measure of the region's escape from the desiccation that has hit most of the rest of Africa.

The south-western Cape is also home to a single species of very primitive moth. Swift moths, Hepialidae, are mainly found in Australia where the survival of these very large-winged insects may be linked with their caterpillars exploiting an unusual source of food, roots. The larvae of the Cape Venus or silver-spotted ghost moth, *Leto venus*, are even more specialized in being wood-borers; their main food is the timber of the Kuerboom trees. Such large, vulnerable and primitive moths are thought to be Gondwanan relics that have survived in relatively stable corners of the former mega-continent. In the eastern Cape up to Maputo is another localized endemic, the elegant Comma moth, *Spiramiopsis comma*. This is not a primitive species and, as with other moths, its limited range could be linked with a food plant, in this case a species of *Secamone*. The butterfly fauna of southern Africa resembles the proteas and heaths, in that limited stocks of relatively ancient butterflies have proliferated to produce large numbers of species, each with a very small range that does not overlap with related neighbours. The most spectacular radiations are in coppers, *Phasis, Thester, Dira* and *Aloeides*. The protea coppers, *Capys*, are typically associated with protea flowers and the few tropical species are found in uplands where proteas grow (in

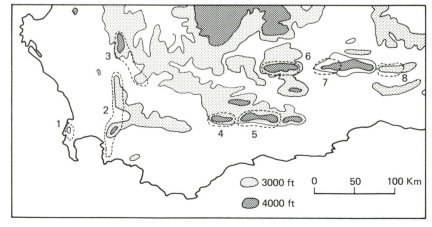

Cameroon and the east African mountains). Endemics of obviously tropical affinities are few but do occur, notably *Papilio euphranor*, *Charaxes karkloof* and *C. pelias* (the latter is attracted to protea flowers but uses another southern endemic, *Schotia*, as its food plant).

Cape highland heathers provide food for another group of interesting endemics. Stag beetles of the genus *Colophon* are mostly restricted to hill-tops within the winter rainfall area (where they first caught the attention of hikers and several are named after their mountaineering discoverers). Few of the twelve species overlap in range (see map). The beetles are unable to fly and are further limited by their larval diet of heather roots. Climate fluctuations have evidently caused the spread and subsequent isolation of populations within this minuscule realm, but what of their broader affinities? Stag beetles are entirely absent from the tropics and the most advanced species are northern temperate insects. The Cape beetles are primitive and have not elaborated 'antlers', but stag beetles are known from deposits of 50 million years ago and similar types to the Cape species live in Australia and Chile. Unlike the cycads, stag beetles are probably not sufficiently ancient to be actual Gondwana relics, which implies passage across the tropical regions they no longer inhabit. Rather than a chance traverse it is more likely that most such southern isolates derive from formerly widespread and versatile groups that have since declined or become more specialized. Depending on the animals or plants involved, the periods of such wide distribution could range between less than a million and several tens of millions of years ago.

Many puzzling anomalies in biogeography are partly explained by the progressive restriction of older forms to more difficult and outlying zones where they may become more or less specialized. It is a pattern that suggests itself for many plants and animals. This is also not inconsistent with a more limited regional theatre in which ancestral populations first evolved or flourished in the southern half of the continent only to become progressively more and more constricted. This may have happened to a genus of chameleons, *Bradypodium*, which has fourteen species confined to South Africa. Their original range may well have contracted under pressure of competition from the dominant tropical genus *Chameleo*. (One of these endemics, *B. pumilus*, is depicted on page 72.) Among other reptiles restricted to this area are ten species of tortoises including the beautifully patterned Geometric Tortoise *Psammobates geometricus*, which is nearly extinct.

In the Rocky Mountains of America the cold torrents are inhabited by frogs that can cling to rocks in rushing water and the tadpoles of which have modified their mouths into suckers. In the Himalayas a local frog has the same habit. The frosty heights of Table Mountain and other massifs of the Cape give rise to short streams that rush down their steep sides, often within sunless gorges. Here the very primitive Ghost Frog, *Heleophryne*, has found similar solutions to those of the American and Asiatic frogs in overcoming the problems of living in rushing water. The tip of each toe in adults is reminiscent of an octopus sucker, while thighs, belly and throat are capable of being squashed flat against the slimy surfaces of its habitat. The skin in these areas has patches of minute hooks like Velcro. The tadpole is even more extraordinary; its mouth is broader than the body, which has become a mere appendage to the mouth and tapers away into a long tail. Not only can the tadpole cling to rocks in rushing water but it can graze off the algae that smother them.

Suckered toes are typical of tree frogs and are present in another well-known Cape species, this time a neoendemic belonging to a very common widespread genus. The Arum Frog, *Hyperolius horstocki*, has earned its name by the frequency with which it is found lying at the bottom of the African Arum Lily's tapered cone. There it waits for insects, which are seized by its sticky tongue. The frog,

Geometric Tortoise, *Psammobates geometricus*.

Speckled Cape Tortoise, *Homopus signatus*.

Cape Ghost Frog, *Heleophryne rosei*, (an archaeo-endemic).

Arum Frog, *Hyperolius horstocki* (a neo-endemic).

Cape Sugar Bird, *Promerops*.

Ground woodpecker, *Geocolaptes olivaceus*.

like a chameleon, can change colour. In the ivory recesses of the arum or in bright sunlight it can become almost white, and the frogs actively seek white objects during the arum's flowering season.

At night, against a dark background and outside the flowering season the frogs are dark brown with green spots. Dark and light bands down its sides turn green, yellow, pale grey or brown. When the light is dappled the frog's skin colour responds by matching the sunlit and shaded areas.

Unique flora generate many peculiar associations with animals. For example, numerous insects have adapted to the proteas; one, the Protea Beetle, feeds on nectar and is also an agent in the pollination of some species. However, the brightly coloured bracts which enclose an inflorescence of densely massed nectar-bearing flowers have probably evolved to attract birds. The little metallic sunbirds are major pollinators all over Africa, but in the Cape the abundance of nectar in proteas (supplemented by insects) supports the much larger sugar birds *Promerops* spp. So abundant is the nectar that it is boiled down to syrup in at least one species, *P. repens*, hence the local name sugarbush. Proteas have Australian relatives, banksias, silk oaks and macadamias, which also secrete copious nectar and these are visited by nectar birds that have been given the Latin name Meliphagidae or honey-eaters. While some ornithologists link them with starlings, others have allocated sugarbirds to the otherwise exclusively Australian family. Several explanations for this peculiar distribution have been suggested: 1. Flight across the Indian Ocean (today they seldom fly more than 600 m). 2. Convergent evolution by two unrelated birds. 3. Pan-Indian Ocean distribution broken by extinction. 4. Pre-adaptation by a common ancestral stock of birds to nectar feeding at a time when the birds were more mobile and the southern continents were closer than they are now. The last option becomes less unlikely when it is remembered that 70 million years is half the time that birds have been around and that nectar-sucking (with a hollow-tubed and brush-tipped tongue) is such a peculiar way of earning a living that there is little scope for departure from type, only for the acquisition of ever more refined techniques. Nectar feeders like ant or termite eaters require such elaborate anatomical and physiological specializations that the earlier the necessary transformations began, the more incontestable the niche became. This is a truism for many groups of animals.

This presumably ancient association can be contrasted with a 'modern' one. Proteas support a canary which not only takes nectar, flower parts and insects from the protea but also protea seeds as its main food. Compared with the sugarbird the Protea Canary, *Serinus leucopterus*, is a very generalized feeder. When the likely period of its association with proteas is compared with that of the sugarbird it is very much of a newcomer. It is only when compared with other canaries that it too becomes a relict and specialist. Four or five species of canary are widespread across Africa, living in all manner of habitats (several also visit protea bushes but are less dependent on them). The only close relatives of the protea canary are restricted today to the Atlantic Benin Islands and the high mountains of eastern Africa, where they live in heathland and along the margins of forests. The inescapable inference is that the commoner species (possessing more powerful beaks, or else smaller and more agile) have spread at the expense of older types of canary, now represented by the island relics.

The bird fauna of southern Africa changes radically with the disappearance of forest west of Knysna. Here is the main stronghold of the Knysna Touraco, the southern isolate of a mainly forest and woodland group (see plate on p. 178). The larger insect-eating species are also scarce in the extreme south-west. However, grasslands and open areas on the highveld support a large and very unusual ground-dwelling woodpecker, *Geocolaptes olivaceus*, which subsists on ants and termites and nests in holes that it digs out of earthbanks. This is a semi-social species and it is uncertain as to whether it is a relative of similar woodpeckers in Eurasia and South America or a convergent species that derives from tropical African woodpeckers.

'X ray' image of a Giant Golden Mole, *Chrysospalax trevelyani*, digging.

The mammals too have some of their most primitive relicts stranded in southern Africa. Golden moles are blind, subterranean descendants of the very earliest types of mammals. Among the many types found in southern Africa is the Giant Golden Mole *Chrysospalax trevelyani* (which happens to share its habitat with the Cambrian *Peripatopsis*) and helps support its 23 cm long body on a diet of earthworms that are larger here than anywhere else in Africa.

The Cape's role as a refuge for animals that cannot compete with more advanced tropical relatives is well illustrated by antelopes. Africa is the land of antelopes: in tropical east Africa there are 48 bovid species, 14 of which are antelopes less than 20 kg in weight. Southern Africa has 8 small types out of a total of 30 species, one of which is extinct and five of which are endemic to the Cape. Antelopes with their many ecological and dietary specializations all derive from smaller short-horned ancestors that are thought to have had less complex societies and less difficult diets. This radiation has taken some 20 million years or more to develop. The modern antelope that most closely approximates to the model of a primitive common ancestor is the Cape Grysbok, *Raphicerus melanotis*. Only the rock-dwelling Klipspringer and the fruit-eating Common and Blue Duikers share any parts of the grysbok's habitat in the western Cape. Other small antelopes, Suni, Red Duiker and Dik-Dik are not far to the north but there is no overlap, so the Cape Grysbok has the foreshore and protea scrub inland very much to itself.

The Cape Grysbok, *Raphicerus melanotis*.

Another somewhat less primitive antelope, the Rhebok, *Pelea capreolus*, inhabits the cold uplands of the Cape and highveld hinterland. Its Afrikaans name derives from Dutch settlers likening it to the roebuck they had left behind in Holland. Its scientific name is a latinized version of the Tswana name Phele. The anatomy and habits of this slender grazer ally it to two quite distinct tribes of antelopes: the dwarf antelopes or Neotragines and the reedbucks or Reduncines, lineages that may have parted 13 million years ago. Just how the rhebok relates to that divergence is still unclear, partly because the animal has continued to evolve within its southern enclave. For example, few antelopes in Africa have really dense, woolly coats and it is some measure of its long tenancy of these cold highlands that the rhebok has a thick grey fleece.

The sugarbird and the grysbok are by implication ancient residents of the Cape but there is also the antelope equivalent of the Protea Canary—a more recently arrived species already made obsolescent by rapid evolutionary change in the tropics. The duller-coloured Blesbok, *Damaliscus dorcas phillipsi*, once inhabited the extensive highveld grasslands behind the Drakensberg range, while the coastal flats around the Cape swarmed with herds of glossy piebald animals that the Dutch settlers called Bontebok, *Damaliscus dorcas dorcas*. The hartebeest family appear very late in the fossil record, only becoming numerous and varied within the last 2 million years. The hartebeest trend has been towards larger size, greater mobility, larger herds or heavier, more complex horns; Topi *Damaliscus lunatus*, hartebeest *Alcephalus* and Brindled Gnu *Connochaetes taurinus* have developed these attributes and they dominate the whole of savannah Africa. The Bontebok derives from an earlier stage in a very recent evolutionary radiation.

Vaal Rhebuck, *Pelea capreolus*.

The bonte- or blesboks are to the hartebeest what the White-tailed Gnu is to the Brindled Gnu. The White-tailed Gnu *Connochaetes gnu* is a diminutive caricature of a bison, its lashing tail, face and chest tufts and mane promote the illusion of size and weight to an otherwise insignificant beast. In the Pleistocene period this type of antelope was widespread, even in tropical Africa. At the time of European colonization it was extremely abundant in open grassland from the

Bontebok, *Damaliscus dorcas*.

White-tailed gnu, *Connochaetes gnu*.

north-eastern Cape province to Natal and southern Transvaal. In 1836 Cornwallis Harris wrote:

De Jagt van de wilde beest (gnu-hunting) forms a favourite diversion of the Dutch colonists and occupies a very large proportion of the apparently valueless time of the trek-boors or nomad farmers who graze their overgrown flocks and herds on the verdant meadows . . . in a single heap I have seen so many as two or three hundred mouldering skulls and . . . from the numerous skeletons everywhere strewed the mortality among the species must be very great.

Soon the White-tailed Gnu was extinct as a wild animal but a few landowners, including mining interests, permitted some gnu to remain on their estates. Even so by 1947 only three small herds could be found in the entire Cape province. By the 1970s widespread concern for the destruction of South Africa's indigenous flora and fauna had begun to exert pressure on the authorities, and the White-Tailed Gnu now number several thousands scattered through many reserves and farms.

In the Cape there has been more lost, largely because there was more to lose, than in any other part of Africa. One of the first casualties was another antelope, the Bluebuck, *Hippotragus leucophaeus*. In the structure of its teeth and horns this species was intermediate between the Sable and Roan but was smaller than either of them. A contemporary hunter described their coat as looking like 'blue velvet in life but fading to lead colour in death'. Even more faded are today's relics: four ill-mounted specimens, a skull and a couple of amateur watercolours.

The extinction of the Bluebuck has sometimes been cited as inevitable, the final push at an animal teetering on the brink of natural extinction. This is to oversimplify things; Roan, Sable and oryx antelopes have all seen rapid decline because their behaviour renders them particularly vulnerable to hunters. They make no attempt to hide, they run fast but are without stamina, have conservative and predictable habits and live in small groups on land to which they are strongly attached. By all accounts the Bluebuck conformed to this pattern and local archaeology suggests that their range may have slowly contracted over many centuries of hunting and settlement by Khoisan people, whose numbers in the seventeenth century have been estimated at 200,000 south of the Orange River. In spite of such a large human population the foothills and lowlands of the Cape would have remained a relatively stable refuge for Bluebuck and Bontebok, because the native Chochoqua were primarily herders and traders (indeed by 1614 at least one Chochoqua was in London to learn the English language and European trading practices).

The fate of both Chochoqua and the Bluebuck was ultimately at the mercy of the accelerating trade between Europe and eastern Asia. The single most influential event took place in an Amsterdam boardroom of the Dutch East India Company in 1650. This meeting decided to set up a 'Tavern of the Seas' beside a sheltered stream flowing off Table Mountain. This 'tavern' was to offer locally grown vegetables, flour, salted meat, water and eventually wine and medical attention to the ever-increasing number of Dutch, British, French and Portuguese ships in need of revictualling. The tavern competed with and soon displaced local Khoisan, who for over a century had traded livestock, shellfish, ivory and other local products for beads and metal goods with passing ships. Within 25 years the expanding settlement had become the fiefdom of a Dutch-Indonesian father and son, Simon and Willem van der Stel, who were governors of the Cape of Good Hope from 1679 to 1707. Ever larger numbers of ships needing ever larger quantities of food had led to the conquest of the Chochoqua and to the arrival of some 2000 Europeans, who were induced to come and farm the land that had been seized. This area included all the remaining range of the

Bluebuck and Bontebok. The rapacity and corruption of the Van der Stels led to the eventual dismissal of Willem as Governor but their long reign had set in motion and sanctioned habits of exploitation of man and nature that are still in evidence.

The Bluebuck, *Hippotragus leucophaeus.*

This period saw the main decline of the Bluebuck, although the last animals only died in 1779. The Bontebok were more fortunate and owe their rescue from extinction to a single family, the Van der Byls. These enlightened farmers had set aside a part of their estate 'Nacht Wacht' for the Bontebok as early as 1837. Numbers at that time were down to 27. Neighbouring farmers followed their example but it took 100 years before a small national park was set up south of Swellendam. About 1000 animals now live in the fenced Bontebok National Park and on other Cape reserves and farms.

Efforts to conserve plants and animals in the Cape go back a long time. The forests around Plettenberg Bay were appropriated by the British Royal Navy in 1811 and its self-interest was behind the first Reservation Ordinance published in 1846. Appointment of 'The Cape Botanist', Ludwig Pape (under the influence of Sir William Hooker in London), followed in 1858. Unfortunately Pape died in 1862, the year of publication for his 'Silva Capensis' (which made the first real call for scientific conservation). Pape and the Cape Governor, R. Rawson, had drawn up a Forest and Herb Preservation Act in 1859. However, these gains were lost in 1886 when the settler community and loggers combined to oust Pape's successor, J. C. Brown, who was forced to retire.

The idea of a botanical garden was first mooted by the Swellendam Agricultural Society in 1845, but it was only in 1913 when the Eksteen hunting estate and lodge, called Kirstenbosch, came on to the market that a young professor of botany at Cape Town University succeeded in founding a National Botanical Gardens. The garden's 500 hectares merge with the steep eastern slopes of Table Mountain to serve as much as a natural reserve as a garden for indigenous plants. Pearson's example led to the formation of five other regional botanical gardens. Kirstenbosch is now one of the finest botanical centres in the world and the university continues to be a focus for conservation. A plant survey group led by Dr Tony Hall at the university is currently struggling to save some of the 1450 species of rare or threatened plants from an avalanche of development projects and in the face of lukewarm official support. Even so, Dr Hall and his colleagues have surveyed and helped to set up 10 of the 19 major plant sanctuaries that would be necessary to ensure some sort of future for the majority of Cape flora. They have also mobilized weekenders and volunteer enthusiasts to try to root out or control exotics such as the Australian Wattle that have invaded stands of rare indigenous trees.

The Cape and its long eastern shores are especially attractive to farmers and South Africa's human population is growing by 2000 people a day, some three-quarters of a million a year. Edible or money-earning plants are steadily replacing the Cape flora. It is a pity that gardeners, when they buy a flower or packet of seeds cannot pay a small royalty towards the health and well-being of the plant's original home. If they did, Cape endemics would earn massive royalties, but plants, unlike musicians or authors, do not have unions and lawyers.

The great ecologist Aldo Leopold remarked that one of the penalties of an ecological education is that one lives in a world of wounds. Nowhere is this more true than in the Cape. Here is a landscape so bloodied that little can be seen of the once-healthy body. The largest patch of relict vegetation in the Cape lowlands is 750 hectares, and among the 28 endangered plants found in this single site there are several now known from nowhere else. Thirty-eight species have recently become extinct, 68 are in grave danger and more than 1000 are suspected to be at risk.

It can be argued that natural processes will eventually extinguish many endemics, especially uncompetitive older types, and that conservation may simply interfere in or delay this process. For example, the Cape Heather, *Erica farii*, is naturally confined to a one-hectare patch close to Cape Town.

It is probably just as important to try to understand the processes that have brought a species to such a pass as giving preference to the conservation of man-threatened species. However, it is as much moral as scientific arguments that make the latter more compelling. In the Cape the distinction between natural and human-induced extinction is rendered almost meaningless by the dense mosaic of cultivation, roads and settlement. Alien invasions and diseases, changed grazing and firing patterns, altered water-tables, the loss of pollinators and dispersers, air, soil and water pollution all exert their influence and are probably hastening many species to extinction. In the face of these vicissitudes highly refined specialists are just as likely to suffer as declining 'has-beens'.

A major turning-point in public awareness began in 1968 when Dr Anton Rupert founded the South African Nature Foundation. Since then, persistent pressure from SANF has resulted in the inauguration of many new national parks and the setting up of a National Heritage Scheme which helps landowners and local conservation bodies to voluntarily set up their own private reserves.

The conservation of wild plants and animals often depends on the goodwill of farmers and an unusual aspect of conservation in South Africa is the popularity of local flower and agricultural shows, particularly those in key localities such as Caledon and Hermanus. At these shows many farmers exhibit wild flowers that grow on their land and so have the incentive to save areas from the plough. These shows have been strongly supported by the Botanic Society of South Africa and the National Botanic Gardens. These and some provincial bodies, notably the Ramskop Reserve, not only preserve local endemics but try to take pressure off collection from the wild by selling the seeds and seedlings of rare plants for horticulturalists: proteas, bulbs and cycads are among the most popular. In the last 10 years more than 30,000 cycads have been sold to the public from one nursery alone.

In 1983 the National Plants Campaign was launched with a wild flower show, 'Flora', in Cape Town that is now an annual event. The South African Nature Foundation organized this campaign with support from the Botanic Society and the National Botanic Gardens and used the event, amongst other means, to raise funds to buy land for conservation, especially in areas of high plant endemism.

SANF also supports a weekly Nature Supplement called 'Our Living World' which accompanies major Sunday newspapers. The Cape flora therefore has influential friends. Sadly, there are even more powerful forces threatening it. The Government's Armaments Corporation, Armscor, has plans to develop a missile range in the Outeniqua Mountains which threatens the De Hoop Provincial Nature Reserve, a major reserve for 'fynbos'. Fynbos now covers less than 1% of South Africa, yet 65% of the threatened endemic plants occur in this type of vegetation.

The conservation of a particular plant, animal or place has been assisted in South Africa as elsewhere by appeals to sentiment, patriotism and tradition. The largest single grove of Modjadi cycads (with cones weighing 36 kg) grows in a misty valley of the Letaba River known as Devil's Gorge. Its survival is due to prohibitions on interference from the Rain Queen, a traditional leader among the local Venda people. By contrast the less spectacular cycad, *Encephalartos woodii*, is now extinct as a wild plant.

The declaration in 1976 that the Giant Protea, *P. pycnaroides*, was to be the national flower showed official endorsement for a growing public awareness of the country's unique flora. Notwithstanding this, the Blue-bearded Iris, *Moraea loubseri*, a flower as beautiful as the protea (see colour plate, p. 71) and only discovered in 1973, was well on the way to being literally blasted out of existence a few years later, because contractors of Langebaan were quarrying the little granite hill to which it was restricted. Fortunately this has been halted and the plant grows well from seed and is being propagated as a garden flower.

In spite of several centuries of rampant destruction having reduced much of its natural richness to vestiges and in spite of the economic and political power of vested interests promoting still further destruction, South Africa has an active and influential conservation lobby and a large network of small reserves and parks. (See Appendix.)

From small beginnings, keen and for the most part amateur naturalists in southern Africa have saved from extinction the Bontebok, White-tailed Gnu, Southern Grass Rhino and many less spectacular animals and plants. Above all, they have set aside and continue to press for areas where whole natural communities can survive. These efforts have been made in the face of powerful pressures which, as the missile range plan shows, are still increasing and still need to be opposed.

A grove of Modjadi cycads,
Encephalartos transvenosus.

CHAPTER 3

Skeleton Coast, Diamond Coast

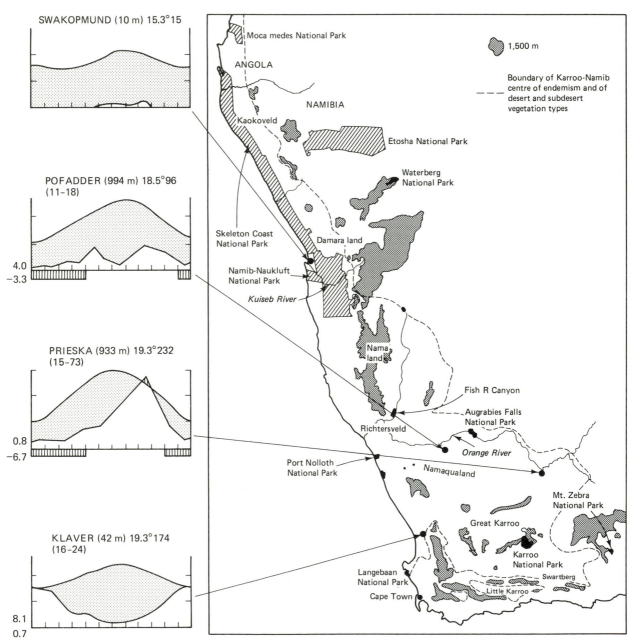

SWAKOPMUND (10 m) 15.3°15

POFADDER (994 m) 18.5°96
(11-18)

4.0
-3.3

PRIESKA (933 m) 19.3°232
(15-73)

0.8
-6.7

KLAVER (42 m) 19.3°174
(16-24)

8.1
0.7

Moca medes National Park

ANGOLA

NAMIBIA

Kaokoveld

Etosha National Park

Waterberg
National Park

Skeleton Coast
National Park

Damara land

Namib-Naukluft
National Park

Kuiseb River

Nama
land

Fish R Canyon

Augrabies Falls
National Park

Richtersveld

Orange River

Port Nolloth
National Park

Namaqualand

Mt. Zebra
National Park

Great Karroo

Karroo
National Park

Swartberg

Langebaan
National Park

Little Karroo

Cape Town

1,500 m

Boundary of Karroo-Namib
centre of endemism and of
desert and subdesert
vegetation types

Arid zone of south western Africa showing boundary of Karroo/Namib centre of
endemism, bordering uplands and National Parks.

Few landlubbers think of oceans as places of drought, still less imagine that cold sea air can stay heavy with fog without ever falling as rain; yet a great part of the south-eastern Atlantic is just such an oceanic desert. What is more, it is a desert that happens to extend over land. This huge ellipse of drought has an abrupt eastern margin hard up against the Kaokoveld escarpment in Namibia. Towards the south the dryness obtrudes further and further inland yet just as temperatures drop towards the Cape the desert air's influence is softened as it buffets against warmer winds bearing moisture from the Indian Ocean.

Although the temperate Cape coast is touched by this meeting of extremes it is the mountains inland that are the great divide. Beyond these fringing heights harsh Atlantic drought grips a land that the Khoisan herders and hunters called 'Karroo' or 'Garob', the place of great dryness. Studded with eroded flat-topped mountains, it is a plateau that falls away into the vast flat valley of the Orange River. A thin khaki scrub is scattered across mountains, gravel plains and sandy valleys. It is a type of vegetation that stretches westwards to the sea in Namaqualand and north in an ever-narrowing strip to Angola, 2000 km away.

North of the Orange River lies the largest official 'out-of-bounds' in the world, 60,000 sq km of 'Verbode Gebied'. The pretext for this grotesque prohibition is that the long desert foreshore is littered with diamonds. This barren diamond coast is an apt metaphor for the flora and fauna of the Karroo and Namibia deserts. Treasure is concealed here beneath an exterior that is only made drab by distance and drought. Of some 4000 species of plants more than half are endemic, their dominant common adaptation is the ability to wait.

One example is a spiny thick-trunked plant called *Pachypodium* or 'elephant's foot', which can germinate during a storm and send down a root in its first 24 hours of growth. It may then sit, hardly growing for as many months or years. Because these plants rely on very rare periods of sustained rain to make real progress, they tend to form single age stands. At about 2 m in height, these thick-bodied, spike-crowned plants resemble human figures standing in posed groups. The Khoisan Nama call them 'half men' and they have built them into their own mythology. The story goes that the Nama migrated south long ago with the injunction never to look back. Since the 'half-men' tilt their 'heads' to face the northerly sun, and thus shade their trunks, the Nama identify them as ancestors punished by God for looking back.

Other examples of biological patience are provided by the gladioli, which may wait 10 years to flower, and a dishevelled-looking corpse called the Resurrection Plant *Myrothamnus flabellifolia*. This plant only needs a few showers to activate chlorophyll, which soon flushes its shrivelled grey leaves into greenness and life.

These plants await the rare triumph of the Indian over the Atlantic Ocean. Every now and then, sometimes decades apart, two events coincide—the persistent skin of cold dry winter air retreats seawards while a succession of rain clouds blow in from the east and drop their burden on the thirsty land. Suddenly every plant and animal must make the most of it—everything uses the water to reproduce, and reproduction involves action and sets off lots of competition. Noisy birds, frogs and insects sing, croak and chirrup while plants generate their own brand of 'noise' in the colours and scents which they broadcast to attract pollinators. What the rare winter deluges set off is a huge advertising convention where the exhibition stands are mostly flowers and the visitors are insects (plus a few birds). A winter flowering in the Karroo lasts no more than a few weeks, yet there is no other event to which it can be compared; it is one of the wonders of the world. Valleys, flats, hillsides and mountains explode in colours that are as intense as they are fragile and ephemeral. The succulents and mesems pack their purple or white blossoms into dense canopies that smother every parent bush. Out of the sand spring daisies that mass their sunny flowers in solid

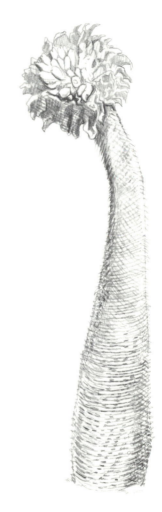

Halfmen, *Pachypodium namaquanum*.

carpets of yellow, orange, white and blue. Aloes and lilies throw up elaborate candelabra or trumpets of red, orange, pink and yellow. There are patches of wild flax and irises.

Many of these Karroo plants, especially the daisies, lilies, mesems and other succulents, are easy to propagate and were collected for seed merchants more than a century ago. Miniature showings of this greatest of all wild-flower spectaculars were long ago transplanted to gardens all over the world. Few of the gardeners will know of the strange quirk of geology, climate and geographic isolation that generated such exquisite beauty of colour and such variety of form. Among the puzzlers is the Beetle Daisy, *Gorteria*. Short with yellow-hearted flowers, each petal, orange above and purple below, carries a bold black spot which seems to mimic a beetle. These flowers are less damaged by beetles than neighbouring daisies, but why this should be so remains unknown. What is clear is that while all Karroo plants must wait for the rain the waiting can be done as a seed, a root or a dormant plant.

Weltwitschia, *Welwitschia mirabilis.*

Seeds, being particularly vulnerable, must be mass-produced for survival, yet are very 'cheap' in materials. To grow fast, attract the necessary pollinators and then to broadcast billions of feather-light seeds upon the winds before dying has been the great triumph of daisies. The succulents survive by storing water in their leaves and stems; the lilies, irises and aloes by storing it in their bulbs.

One of the most primitive and extraordinary of Namib plants is the Welwitschia. *Welwitschia mirabilis* consists of a deep tap root, a very short spongy stem and a pair of long strap-like leaves that continue growing at the rate of about 10 cm a year over the 2000 years or so that a plant is thought to live. The leaves slowly wear away at the tips but fog water condenses on their living surfaces which are covered in stomata. The water enters these m   : valved perforations and is rapidly transported to the stem. In a wind or under dry conditions the stomata close and thus prevent precious moisture from escaping. Male and female plants grow rather similar red cones that emerge between the leaf bases. Pollinated by desert insects they produce winged seeds that germinate in the rare wet years. Welwitschias occur all down the length of Namibia with occasional clusters where they may be relatively common (notably the plain between the confluence of the Swakop and Khan Rivers, close to the Rossing mine complex) but they are generally rare and widely scattered.

Quiver tree, *Aloe dichotoma.*

The habitats in which the Namib and Namaqua flowers grow are varied: sand, gravel and rock (there is little in the way of true soil) and each carries different communities. Furthermore the climate in south-west Africa changes scarcely at all, or very slowly, from south to north but very rapidly from west to east.

Along the surf-lashed seashore the sand is white. Inland it turns from pink to orange and along the driest, coldest and windiest section of the coast it piles up into the largest dunes on earth. They run parallel to the sea for 600 km, some standing more than 400 m high and up to 2.5 km wide. All this sand has been eroded out of the inland plateaux over a million or more years and there are large stretches of dunes where no vegetation grows at all. A few seasonal rivers cut through the deserts, flowing out from narrow irregular chains of hills and mountains inland. The coast itself is cold and foggy. Well inland, the desert steppes are hot and dry. In between lies a zone where it scarcely ever rains, where freezing nights are followed by grey, fog-bound dawns and dusks and a scorching midday.

Fish River canyon.

The larger animals are all resistant to drought but are mobile. Oryx, zebras, Brown Hyenas and jackals move easily across the narrow north–south bands but the smaller animals are more strictly confined. The dunes conceal a golden mole, *Eremitalpa granti*, which 'swims' in the loose sand using its broad hollowed claws. This mole feeds on beetles, spiders, termites, ants and lizards, which are probably caught mainly on the surface. Here in turn the mole is vulnerable to owls and its bones and fur are frequently found in the pellets of local owls. How the golden mole breathes when encased in sand remains an enigma. This very specialized insectivore is found in dunes where *Stipagrostis* and *Trianthema* grasses succeed in retaining some of the fog that rolls over the dunes every night. Not only can the leaves absorb moisture and send it to the roots but the roots in turn radiate out to 10 m, just below the surface, where they can pick up fog condensing on the sand.

The Atlantic fogs have given rise to a whole dune-living fauna and flora found nowhere else on earth, all of which extract water by a variety of ingenious methods. The most successful and varied of these are the toktokkie beetles, which feed mainly on wind-blown grass and grass seeds. One of them, *Onymacris unguicularis*, up-ends itself along the crests and dunes and lets the dew run down its carrunculated coal-black body into its mouth. Another beetle, *Lepidochora*, throws up ridges along the sides of a metre-long trench. Built across the driftline of the fog, these ridges catch significantly more dew than the surrounding sand and, as the beetle retraces its track, it sucks up this meagre harvest of water as it goes. Plants also rely on fog condensing on their bodies; one white, papery plant locally known as 'baboon fingers', *Anacampseros*), absorbs dew from the droplets that condense and trickle down its swollen leaf-fingers.

Two kinds of lizards live in the dunes. The larger, Shovel-nosed Lizard, *Aporosaura*, is a versatile feeder on grass and insects. This agile fringe-footed animal frequently interrupts its sprints over the surface with sudden shovellings with its nose in the sand. Possibly there are scent or temperature clues to be picked up in the sand, but the shovelling is also a form of signal to other lizards,

Namib beetles.
Centre and left, Dune beetle *Onymacris unguicularis* collecting fog droplets.
Top right, red and yellow beetle, *Cardiosis fairmairei. Right* a discoid trench beetle *Lepidochora* sp.

Namib reptiles.

for males have been seen to perform these ritualized movements during courtship. Whatever the functions involved, the structure of this lizard's head has been modified to form an elegant reptilian scoop. Another common lizard is a tiny web-footed species called *Palmatogecko* which feeds on termites. Like the dune-top beetle this gecko allows fog to collect on its skin, and its long tongue is specially adapted to slither all over its face sweeping the precious liquid into its mouth. This is a pretty nocturnal species with a red back, yellow sides and enormous slit-centred black eyes.

Another dune-living reptile is the little Side-winding Adder, *Bitis peringueyi*, which hides its 25 cm body in the sand, leaving only its eyes and tail tip above the surface. When a lizard or gerbil appears the tail twitches and so serves to lure the snake's prey within striking distance.

One of the few birds permanently resident within the sand dunes is Gray's Lark *Ammomanes grayi*. The bird's plumage is perfect camouflage for concealment against the white or grey gravel flats to which it is strictly confined. Apart from an ephemeral growth of thin grass, the larks have virtually no shelter from the kestrels and chanting goshawks that are not uncommon in the Namib. Being drought resistant, having a versatile seed and insect diet and being capable of colour variation seems to be the secret of their very local success. It may be significant that this lark is very closely related to the Desert Lark of North Africa and the Middle East, regional populations of which are particularly well matched with their local soil colours. Having colonized the south during some very arid periods these larks are probably unable to compete with the multiplicity of other larks in less harsh habitats. The fringing sandveld of the Kalahari supports nine species of larks.

Colour matching has also been the special facility of another 'set' of regional endemics. The edges of the desert are occupied by the very pale and elegant Ruppell's Bustard, *Eupodotis rueppelli*; these birds move into the dunes after rain to take advantage of a sudden increase in easily caught insects and the flush of vegetation. (A local finch-lark, *Eremopterix verticalis*, and the Namaqua Sandgrouse, *Pterocles namaqua*, also fly in to make the most of a short-lived flourish of grass seeds and these too are well camouflaged.)

Sparse food probably explains the thin scatter of Ruppell's Bustard along the margins of the harshest desert, but further inland, a multiplicity of other bustards serve to illustrate the special nature of the Karroo and Namib habitats. South-western Africa is not a uniform swathe of dryness but a biogeographic complex containing several sub-islands. Habitats change, sometimes gradually, often quite abruptly, from west to east and south to north. Within these graduated belts plants and animal communities cluster. Each community has an identifiable focus. South of the great Namib dunes are the sandvelds of Bushmansland. Nearby are the gravel plain communities of Namaqualand. A rocky upland community that extends north and south of the Orange River centres on the Richtersveld with its own crop of endemics. Some of these communities, notably the Karroo and Highveld, are very broad and extensive areas but to the north they become long and narrow bands running parallel to the Skeleton Coast. Within this complex the bustards have speciated, subspeciated (and even 'sub-sub-speciated') with extraordinary profligacy. The bustards are known from fossils aged at 50 million years ago and there can be little doubt that southern Africa has offered open habitats suited to them over all that period and that the south-west has been the major centre for their evolution. Division of this region into very dry coastal west and moister but higher south-east and other subdivisions must have assisted the speciation. Indeed, regional differentiation of bustards can be seen today. These mini-Ostriches have several outstanding advantages on dry open plains. Their prime adaptation is probably physiological, the ability to survive heat and cold without surface

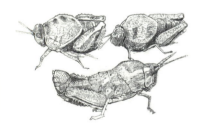

Pamphagid toad grasshoppers.

water. Their superb eyesight penetrates the camouflage of prey and predator equally well (and even plants are camouflaged in south-west Africa). Their own camouflage, especially that of chicks and hens, is without rival and they can crouch, freeze, creep, walk, race or fly according to need. As a group they show exceptional flexibility in their social systems, some species being monogamous, others promiscuous, some making seasonal movements while others are possessive stay-at-homes. Residential habits are made possible by being able to eat almost any class of food, animal or vegetable.

The closest relative of Ruppell's, the Karroo Bustard, *Eupodotis vigorsii*, occurs further south where it is very strictly residential and, unlike the solitary Ruppell's, forms groups of up to eight birds. A local ornithologist has remarked that almost every valley boasts a population with its own tint of plumage. This observation suggests that the immobility of isolated populations combined with selection by predators could have its effect even on such potentially mobile creatures as birds.

These bustards favour the very dry open country that is locally called 'knervlaktse' or 'gnashing teeth' (from the sound of waggon wheels and boots on sharp quartz gravel). Like the stone-like plants that surround them Karroo Bustards mainly rely on camouflage and, as if to compensate for their invisibility, these birds make a very loud nasal trumpeting. Their range overlaps with a much more colourful bird that prefers the open sandveld and highveld further east. The Blue Bustard, *Eupodotis coerulescens*, commonly gathers in small groups and it is possible that several pairs may join up in such groups to defend a joint territory. This species has a loud ringing cry which early hunters took as a challenge to 'knock-me-down, knock-me-down' and it relies more on flight than crypsis to evade enemies.

The range of these species over much of the Kalahari is shared with the small Black Bustard *Eupodotis afra*. This bird is found where sufficient grass or low cover screens its movements but is sufficiently open to allow fast coursing. The males have a finely patterned back and crown but wings, head and neck are tricked out in bold black and white patterns that come into full play during the breeding season. At this time each male perches on a termite mound and grates out a very loud call. It then flies round in a short circle, cranking the wings like punkahs and yelling 'kraak—de wet, de wet, de wet' as it descends in an unstable glide. The purpose of this display is to draw in the secretive, hidden females for mating, after which they disperse and rear their offspring alone.

Ludwig's Bustard, *Neotis ludwigi*, is the most mobile of the endemics. In the deserts and mountains which it inhabits there is insufficient food to support sedentary habits in such a large bird (it can weigh up to 7.5 kg). None the less it confines its movements to the extreme south and south-west and sometimes occurs side by side with its pan-African relative Denham's Bustard, *Neotis denhami*, an even larger bird of up to 12 kg. Mrs Pamela Hall of the British Museum considers that Ludwig's Bustard is the relict of a proto-Denham that was left behind in these southern deserts. At much the same time northern populations would have fragmented into south-Saharan and Somali populations. The former then adapted to savannahs and has subsequently spread widely throughout savannah Africa.

Birds with similar habitat needs and a comparable distribution are the sandgrouse. This group is discussed in more detail in the next chapter where the contemporary ranges of the three species exemplify the 'arid corridor' that has spanned the continent from south-west to north-east during very dry climatic periods. Sandgrouse, bustards, rock hares, oryx and numerous other organisms may have used it like a highway, but the corridor's evolutionary significance is perhaps greatest along its margins. Wherever the route was bordered by rain-fed coasts or highland habitats, narrow serial bands formed that were inter-

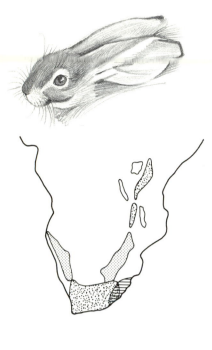

 P. rupestris

 P. randensis

P. crassicaudatus

Distribution of rock-hares *Pronolagus* and drawing of head of *Bunolagus monticularis*.

mediate between moist forest and semi-desert. These would have been narrowest between Lakes Malawi and Tanganyika and close to the east African coast and mountains. A similar narrowing takes place close to the Angolan highlands where the sub-deserts of the south reach up towards their Saharan counterparts. Away from these sharp interfaces, fringing habitats would have been broader but they would have remained serial in nature.

A significant difference between the broader low-lying bands and their 'narrows' on the shoulders of massifs or coasts would have been relative stability in the latter, constant flux and change in the former. With each contraction all grades of arid-adapted communities would have drawn back towards their respective foci in the south-west or north-east (the present pattern). Smaller populations would have remained behind in pockets such as rain-shadows in the lee of great mountains where they have speciated (for example, *Moraea* and *Aloe* in the southern highlands of Tanzania).

In subsequent profiles of the relevant centres of endemism these 'offspring' of arid excursions may be touched upon as elements of the local endemic community (but that is the outcome of a step-by-step regional treatment). Localized species of uncertain origin are scattered along a broad band between the Namib and Somali deserts. The more obviously marginal types tend to occupy quite large ranges in the Kalahari and north-eastern Africa but their satellite relationship to the arid foci can be recognized even though they have ill-defined and less confined ranges. Many mammal species inhabit these margins and an interesting pattern is provided by the rock-hares *Pronolagus*. Jameson's Hare, *P. randensis*, lives in two populations, one centred on Namaqualand and the Kaokoveld, the other in rocky areas of Zimbabwe and the Transvaal. Smith's Hare, *P. rupestris*, also has two populations, one in the Cape and Karroo, the other north of Jameson's Hare, extending as far as central Kenya. A third species, the Short-tailed *P. crassicaudatus*, inhabits the escarpments above the Maputo coast and its adjacent highlands. A still rarer relict hare *Bunolagus monticularis* is slow and clumsy and restricted to a few scattered valley bottoms along the Fish river and in the Karroo.

The mountains and stony outcrops of the Richtersveld and Kaokoveld harbour several relict mammals. The Dassie Rat, *Petromus typicus*, is a member of the very ancient porcupine group of rodents. It lives on intimate terms with the

Ten very localised species of Southern Iris's, *Moraea* spp. in the Cape. The yellow *Moraea longiflora* in Namaqualand, the pale *M. macgregori* north of Klaver, *M. gigandra* north of Piketberg, orange *M. tulbaghensis* at Tulbagh, the purple *M. loubseri* at Langebaan, the blue-eyed *M. villosa* on the Cape peninsula, the spikepetalled *M. worcesterensis* at Worcester, the pink *M. insolens* at Caledon, the polymorphic *M. lurida* around Elim and the pronged *M. barnardi* around Gans Bay.

Overleaf. Bradypodium pumilus, one of the endemic chameleons whose original distribution may well have contracted under pressure of competition from the dominant tropical chameleons, genus *Chameleo.*

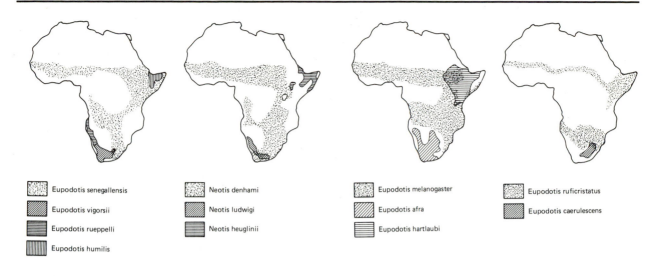

Eupodotis senegallensis	Neotis denhami	Eupodotis melanogaster	Eupodotis ruficristatus
Eupodotis vigorsii	Neotis ludwigi	Eupodotis afra	Eupodotis caerulescens
Eupodotis rueppelli	Neotis heuglinii	Eupodotis hartlaubi	
Eupodotis humilis			

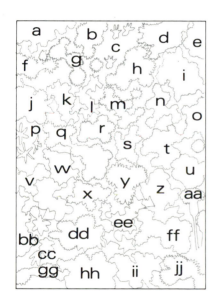

Previous page. Namaqua flowers a *Aptosimum depressum*, b *Dimorphotheca pluvialis*, c *Nemesia ligulata*, d *Aptosimum spinescens*, e & n *Zaluzianskya villosa*, f *Felicia heterophylla*, g *Gazania krebsiana*, h *Arctotis canescens*, i *Sutera ramosissima*, j *Arctotis gumbletonii* k *Heliophila coronopifolia*, l *Sutera tomentosa*, m *Gorteria diffusa*, o *Euryops speciosissimus*, p *Trachyandra tortilis*, q *Osteospermum amplectens*, r *Arctotis laevis*, s *Apatesia maughanii*, t *Ursinia calenduliflora*, u *Drosanthemum speciosum*, v & ii *Grielum humifusum*, w *Pelargonium sericifolium*, x *Gazania lichtensteinii*, y *Dimorphotheca sinuata*, z *Lapeirousia silenoides*, aa *Senecio stapeliiformis*, bb *Crassula hemisphaerica*, cc *Ornithogalum suaveolens*, dd *Carpobrotus quadrifidus*, ee *Wahlenbergia prostrata*, ff *Conicosia pugioniformis*, gg *Frithia pulchra*, hh *Lampranthus comptoni*, jj *Dorotheanthus bellidiformis*, kk *Cotula barbata*.

Opposite page, top. Namib hornbill *Tockus monteiro.*

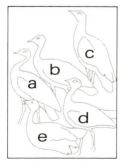

Opposite page, bottom. The Bustards, a *Eupodotis virgorsii*, b *Eupodotis rueppeli*, c *Neotis ludwigi*, d *Eupodotis caerulescens* and e *Eupodotis afra* with maps of their distribution. Photomontage with acknowledgements to C. G. Davis.

much commoner hyraxes (locally known as dassies) but is much more wary. Lizard-like it creeps along on short, spreadeagled limbs and resembles some lizards in readily shedding its dark, bushy tail, which is coarsely textured with an iridescent patch near its base (unlike the body which is a soft, silky brown). To accommodate to narrow crevices it has a broad flattened skull and rubbery ribs. Another endemic rodent, the Pigmy Rockmouse, *Petromyscus collinus*, shows similar flattening to fit into still smaller spaces. Like the Dassie Rat it is a specialized relict.

In the less rigorous habitats of the Karroo and Kalahari there are mammals that have made a niche for themselves in spite of competition from more widespread and successful relatives. For example, the Meercat *Suricata*, with a complex and unique society, is superior here to the Striped Mongoose of tropical Africa. Yellow Mongoose, *Cynictis pencillata*, and the Cape Grey Mongoose, *Herpestes pulvurulenta*, are other south-western regional endemics, the former a social species specializing in a termite and beetle diet, the second the small southern representative of a widespread group that seems to prefer a grasshopper and mouse diet.

The Karroo, Namib and Kalahari are typified by one mammal in particular, the Springbok *Antidorcas marsupialis*. This elegant antelope, superficially so like the gazelles of North Africa and the Middle East, is in fact not closely related. The differences between the two antelope types are subtle but significant and

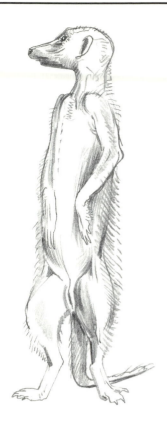

Meercat on guard, *Suricata suricatta*.

Springbok, *Antidorcas marsupialis*, in leaping display.

help to explain why Springbok occupy the gazelle's niche in south-western Africa.

Most gazelles are adapted to heat as well as to drought and their diet is predominantly herbs, leaves or flimsy grasses, as neither their teeth nor their digestive systems can cope with coarse roughage. Springbok, on the other hand, feed off coarse pasture and are adapted more towards cold drought than hot. In northern and eastern African fossil deposits of 2 or 3 million years of age *Antidorcas* are abundant. In the tropics and in North Africa they seem to have been displaced partly by climatic change and partly by the explosion of advanced antelope species that have proliferated there over the last 2 million years. Southern Africa has not only been a continental cul-de-sac but the cooling Benguela current maintains a climate that is probably close to that of past Ice Age deserts in the north. Seen from one perspective the Springbok has been tested and proven a failure in the intensely competitive world of tropical antelope communities. It is, however, impossible to call an animal a failure when it recently ran in millions all over the Karroo and Kalahari. A return of Ice Age conditions might well allow the Springbok to expand once more out of its climatic refuge. Seen from this perspective, areas like the Karroo and Namibia could almost be described as refuges for communities that await (under a geological or climatic time scale) the return of their own special set of conditions.

The point has already been made in the introduction that the distinction between 'old' and 'new' endemics is a relative one. For example, the Mountain Zebra, *Equus zebra*, belongs to a type of horse that (we know from fossils) was once more widespread. It has split into two separate Cape and Namib populations. The former once shared part of its range with the extinct Cape Quagga, *Equus quagga quagga*, which was the Karroo isolate of the highly successful common zebra or 'painted quagga'. Relative to each other the mountain zebra would be more of an 'old' endemic, the Quagga a 'new' endemic. Survival of the first and extinction of the second sheds an interesting light on the process, timing and role of historical accident in extinction.

The Cape Mountain Zebra was already seen to be rare and endangered in the 18th century and an edict was issued banning its hunting in 1742. This was universally ignored and the last big massacre of about 100 animals took place at Craddock in 1910. In 1937 a farm of 1712 hectares was bought as a national park to protect the single mare and five stallions that ran on it. These soon died and little further interest was taken until 1950 when a farmer, Henry Lombard, donated the eleven animals that ran on his Waterval farm to the Mountain Zebra national park that had long been empty of zebras. In spite of official indifference these animals were carefully tended and their progress documented by the park warden, George Hlomela. With the growth of interest and addition of more farms and animals, the population built up until there were 215 by 1980. There are now several healthy populations of Cape Mountain Zebras (and the larger Hartman's race is still widespread in Namibia).

Not even nominal efforts were made to protect the Cape Quagga, so common was it thought to be. In the 18th century various travellers reported vast herds of Quagga in the Karroo. Even in 1832 Cornwallis Harris found 'interminable' herds south of the Vaal River, although he noted that they had disappeared from the rest of their range. In an enthusiastic promotion of the joys of the chase, larded with biblical quotes, he none the less left one of the few detailed accounts of the period. Although he could see that the fauna he called 'a mine of treasure' would eventually be exhausted, he was unable to escape the hedonistic opportunism of the times. Eighteen-thirty had marked the beginning of several decades of *laissez faire* 'Free Trade' under a weak British administration in Cape Town. Trek waggons returned from the interior loaded with ivory and hides; in less than 30 years Quaggas were extinct. A few animals found their way to Europe and it was in this way that the last geriatric Quagga died in Amsterdam Zoo in 1883.

The Quagga, like the Bluebuck, became extinct during a relatively short period of smash-and-grab anarchy. We will not bring either animal back by pointing out that one or two tolerant landholders, a few less venial governors and merchants (or even a more perceptive Cornwallis Harris) would have been all it needed to save these animals. What can be done is to put pressure on today's governors and merchants, as well as recognize ourselves in Cornwallis Harris. Like him, we are able to luxuriate in the experience of rare animals or plants, whether at first hand, in gardens, or books and television. It is in our response to such privileges that we can most differ. In this there is no shortage of example; take Charles Kock, a visiting Austrian entomologist working on the Namib beetles. Through him the Transvaal Museum began a research project in the Namib desert in the 1950s. In 1963 his infectious enthusiasm persuaded Erich Luebbert, a German industrialist, to build a scientific research institute at Gobabeb which today stands in the centre of the Namib Naukluft Park. Symbolically the park has already encroached upon the vast diamond restricted area. It may be too much to expect diamonds to be nudged very far by beetles, but here is evidence for one small step towards a change in values.

Mountain zebra, *Equus zebra* and Quagga, *Equus quagga*.

CHAPTER 4

The Incense Coast, Horn of Africa

The Sahara is today far and away the largest desert in the world but geologically it is a late arrival and largely a result of the Mediterranean becoming a closed basin. The Sahara is therefore a thick but relatively new westerly limb of a vast body of deserts. In Central Asia and the Middle East the dryness stretches on over high plateaux and mountains where hot summers follow freezing winters. From the southern margin of this Afro-Asian complex a long tongue of aridity probes deep into eastern Africa. At present the tip is not far south of the equator but there have been periods when it joined up with the deserts of Namib and the Kalahari.

During such arid periods the Horn of Africa was at the hub of a vast desert world which extended far to the north and west, to the south and to the east. It is an area that was arid when North Africa was forest. For many of the animals and plants that are able to live beneath these cloudless skies, seas and mountains, however narrow, are decisive barriers. It is therefore no surprise to find that the mammals of Somalia are all essentially African. Some of the birds, however, have other origins, as do many of the plants, and it is a combination of African, Mediterranean, Asiatic and very ancient Gondwanan influences that have shaped the unique ecological communities of the Horn of Africa. There are also locally evolved specialists.

Diverse origins are reflected in the distribution of endemics over three distinct zones within the Horn of Africa. The mountainous north is one wing of the Ethiopian dome (see map) and its ecology has much in common with the highlands of southern Arabia and Ethiopia. It is here that Mediterranean and Asiatic influences find their last outpost in Africa. By contrast the Haud, an inland plateau dominated by deep infertile sands, has its main affinities with south-western Africa. The third region consists of a long strip of low-lying Indian Ocean littoral, the ecological isolation of which is reinforced by infertile sandy soils and a consistently powerful offshore wind.

The desert world.

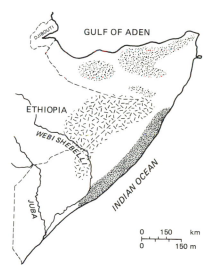

1.	Northern uplands
2.	Haud/Ogaden
3.	Obbia coast

Three sub-centres of endemism in Somalia.

View over the chalk foothills of the Migiurtina range, towards Boosaaso on the Gulf of Aden Coast.

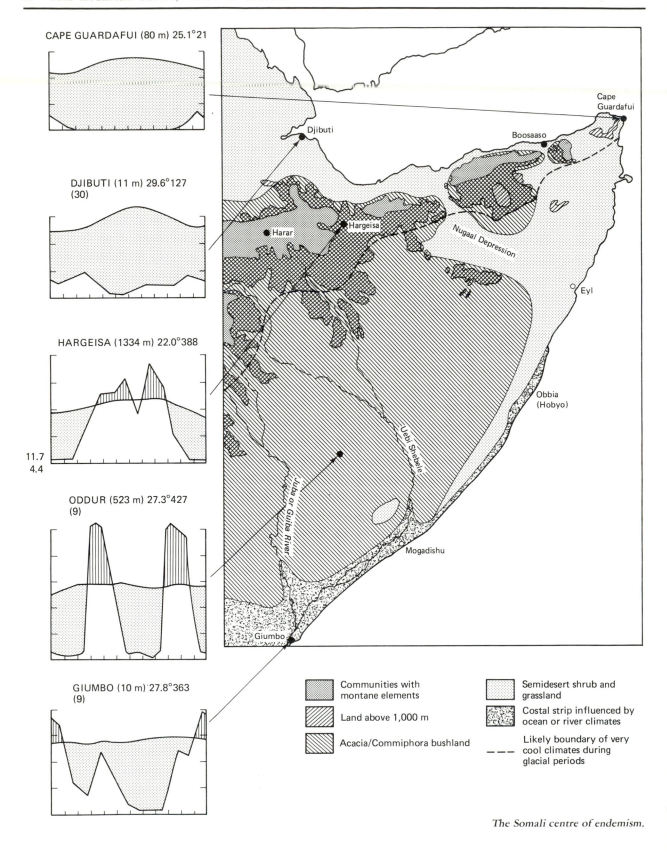

CAPE GUARDAFUI (80 m) 25.1°21

DJIBUTI (11 m) 29.6°127
(30)

HARGEISA (1334 m) 22.0°388

11.7
4.4

ODDUR (523 m) 27.3°427
(9)

GIUMBO (10 m) 27.8°363
(9)

Cape Guardafui

Boosaaso

Djibuti

Harar

Hargeisa

Nugaal Depression

Eyl

Uebi Shebele

Juba or Giuba River

Obbia (Hobyo)

Mogadishu

Giumbo

Communities with montane elements

Land above 1,000 m

Acacia/Commiphora bushland

Semidesert shrub and grassland

Costal strip influenced by ocean or river climates

- - -   Likely boundary of very cool climates during glacial periods

The Somali centre of endemism.

North-eastern Somalia has much in common with Socotra: it shares many unique plants, notably a relative of the geranium, *Dirachma*, and the shrub *Poskea*. This extreme tip of Africa is itself an incipient island, because the Red Sea Rift has continued southwards to cut a broad depression called the Nogal Valley or Dooxo Nugaaleed. The promontory east of Nogal is known as Warsengeli and it is close to its steep north-facing escarpments that some of the rarest and most localized of Somalia's endemics are found. Here a linnet, thrush and pigeon occur as well as a colourful antelope, the Beira. The Nogal depression itself is home to a short-tailed bird called Phillips' Crombec, *Sylvietta phillipsi*. The Nogal depression meets the ocean at Eil, where the low-lying littoral strip, sometimes called the Obbia (Hobyo) Coast, comes to an abrupt end. This point of junction between two biotic subdivisions is a particularly rich focus of endemism. The Horn therefore resembles south-west Africa in having more than one focus of local endemism and, just as the Kalahari and the Karroo support second and third ranks of southern endemics, so too do Somali communities subdivide and adapt to various milder regimes. Today, the 'Somali arid zone' peters out in Tanzania.

The Horn is a very ancient and relatively stable corner of Africa and has an exceptionally rich assembly of arid-adapted species. It is also the Centre of Endemism with the longest history of intensive human occupation. In the latest of innumerable fluctuations the climate of north-eastern Africa has become progressively drier over the last 6000 years. In this time the present population of pastoral Cushitic people has probably had as great an impact on the region as any of the great climatic shifts of the past. After dogs, cattle would have been the first exotic animal to be introduced into the Horn, indeed Cushitic cattle pastoralism is likely to have replaced earlier hunting cultures some time between 8000 and 5000 BC. Camels were domesticated in Arabia nearly 5000 years ago, whence they would have reached Somalia shortly afterwards. Their arrival would not only have extended the range of habitats that people could occupy but would have provided an alternative support in areas that were becoming too dry for cattle. Likewise the sheep and goats that followed, each increased the scope for human occupation but also damaged ecosystems that had evolved in isolation from such voracious herbivores.

Phillips' Crombec, *Sylvietta phillipsi*.

In spite of very intense competition, six very localized antelopes have managed to remain and a larger number of more broadly distributed animals and plants also survive. This is a tribute to their adaptability but there is no doubt about the accelerating overall decline of most indigenous animals and plants. Two of the most localized endemic plants are now listed in the IUCN Red Data Book. One is a conical many-branched euphorbia with white flowers, *Euphorbia cameroni*, the other a squat cuboid succulent called *Whitesloanea crassa*. Three other species with a somewhat larger overall range are also listed.

The Incense Coast or Land of Myrrh of ancient Egypt offered several commodities including ivory but it was frankincense and myrrh that brought the ancients to the two coasts of the Gulf of Aden. As early as 1500 BC fleets of small ships carried many tons of these resins to Egypt. Frankincense is obtained by tapping the bark of *Boswellia* species; the resulting resin provides an aromatic smoke, a scented oil known as Oliben, a sort of chewing gum called Siri and another form which was the old balm of Gilead. In ancient Egypt and Rome these were regarded as perfumes of the gods. Scarcely less valuable was myrrh, a resin extracted from at least four of the 242 known species of *Commiphora* (*C. myrrha*, *C. molmol*, *C. erythraea* and *C. abyssinica*). Myrrh was used as an antiseptic medicine, fumigant, as an oil for anointing living bodies and as an embalmer for dead ones; it was even used in cooking. Some frankincense is still gathered from the *Boswellia* trees which grow on the slopes of Warsengeli because marine mists blow in to condense on dolomitic rocks and thus provide the trees with essential

moisture and minerals. When frankincense and myrrh were valued highly they sustained thriving communities, their towns, ports, ships and camels. The trade also sustained large tracts of woodland in places where all vegetation has a rather tenuous grip on existence. The trees were protected because local people derived wealth and influence from them but ultimately it was the rituals, pleasures and value systems of distant peoples that maintained the incense trade for so many centuries. Now that the trade has become the victim of changing values the question arises—will the goats and fuel needs of impoverished countryfolk eliminate the rare endemics before the value of these plants and animals has been adequately recognized? There is an even bigger threat, commercial charcoal. In Kenya, the rapacity of a powerful political dynasty led to extensive woodlands of frankincense and myrrh being chopped down and converted into charcoal, mostly for export to the Middle East.

The fortunes of at least one Somali endemic illustrate the importance of understanding the ecology of a plant before it can be adequately conserved and its economic potential to local people realized. Much of the hinterland of Somalia is barren and drought-stricken. Like the Nogal depression these lands have helped reinforce the isolation of the Horn but in themselves they have been an evolutionary challenge to various animals and plants. The Yeheb Nut, *Cordeauxia edulis*, is such a plant. The yeheb is a leguminous bush about 2 m in height that only grows on deep infertile sands in the Haud. It bears fatty nuts which taste like sweet chestnuts that have long been an item of food and trade for local people. The nuts and bushes also provide good browsing for livestock (in spite of the red dye in the leaves being so pervasive that it stains the animals' bones and other tissues a bright pink). Onslaughts from livestock are known to have reduced their abundance but the great longevity of the Yeheb (trees are thought to live several hundred years) and its exceptionally deep roots have helped its survival. Natural regeneration of the Yeheb probably depends upon very infrequent periods of sustained rain, because the sand in which they grow should be sufficiently moist for sufficiently long for the tap root to get down and away from the desiccated friable surface. The Yeheb is therefore likely to benefit from artificial propagation (for example, a transplanted population at Voi in Kenya is providing a dye and nuts for the local market). International agencies are now trying to propagate it for cultivation in other parts of the world: how successfully remains to be seen. There is another food plant which also has economic potential but also unsolved problems concerning its germination and propagation. This is a newly discovered locust bean or carob, *Ceratonia oreothamna*. Carob trees were formerly thought to be restricted to the eastern Mediterranean evergreen maquis. The discovery that a second species occurs in the Horn of Africa (and in a single stand in Oman) has served to emphasize the many links that the Somali highlands have with the Mediterranean. Its discovery in 1979 also shows how well-established commercial food plants can be overlooked in corners of the world as remote as the Horn of Africa. Other groups are shared with Arabia such as the Globulariaceae, Cistaceae and Coridaceae.

Another typically Eurasian temperate group that is well represented in Somalia are the lavenders: *Lavendula somaliensis* is one endemic among the seven species found in the Horn. It is peculiar in leaf-form and in re-growing its leaf shoots and flowering heads each year. Powerful aromatics help protect them from attack by insects and livestock alike. Variety of form, colour and scent has made lavenders very popular for herbaceous gardens and they are still used to scent and protect stored clothes in the western world. The Somali Cyclamen is another stranding from Ice-Age Eurasia and the only cyclamen in Africa. Several plants are common to both south-western Africa and Somalia where they have speciated locally, notably *Duvalia*, *Orbea*, *Pentzia*, 28 species of aloes, 80 euphorbias and the succulent stapeliads. The latter are rare but include dis-

Carob *Ceratonia oreothamna*.

Opposite. Somali plants. Top left Yeheb Nut Cordeauxia edulis. Top right Arthrocarpum somalense. Bottom left Somali Lavender Lavendula somaliensis. Bottom right Kissenya spathulata.

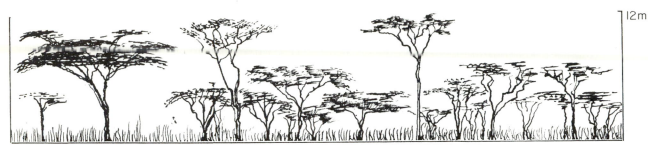

12m

Acacia Savanna   *Acacia* spp.

7m

Hard Pan   *Commiphora, Combretum, Adansonia, Lannea, Acacia Formicaria.*

8m

Thicket   *Commiphora, Combretum, Acacia, Teclea, Maba.*

4m

Semi Desert   *Aloe, Calotropis, Sansevieria, Commiphora, Balanites, Euphorbia.*

tinctive species such as *Huernia somalica*. In spite of its ecological connections with areas as distant as the Namib there is a complete absence of orchids, balsams, ferns and other moisture-loving African plants. This is some indication of how consistently hot and dry the Horn has been.

A third influence, the Indian, is well illustrated by *Dobera glabra*. This species of plant is typical of the Rajasthan desert in India and its scented flowers provide an essential oil that is very popular in India and north-east Africa as a perfume.

Plants from the Obbia Coast show various adaptations to the harsh and peculiar conditions. For example, *Oldenlandia saxifragoides* resists continuous sand-blasting by the wind by growing in a dense cushion-like mound, while *Commiphora planifrons* and *Maytenus obbiadensis* form very low, dense thickets with wind-shaped horizontal branching. Vegetative rooting helps solve the problem of dispersal in high winds and loose sands. Such specialized species represent a fundamentally different type of endemic to those from more representative habitats. In contrast to these local and possibly quite recent developments, the Horn has ancient botanical connections: *Arthrocarpum somalense* is a relict legume found on Socotra and the Warsengeli escarpments with closest affinities to some plants from around the Gulf of Mexico.

There is another American connection in the rock nettle family Loasaceae which has a single old-world representative in the highly localized *Kissenia capensis*. This is a fleshy stinging herb that grows about 80 cm high on pebbly limestone screes and is restricted to the southern Namib, the Horn and Yemen. The rock nettles are evidently both relicts and specialists in Africa, yet a more successful and widespread family in America. There are numerous other examples of distant connections in a flora that has over a thousand plants endemic to the region (and 600 or so that are more narrowly restricted). Birds also reinforce the picture of multiple origins.

Socotra and Warsengeli harbour a very beautiful Golden-winged Grosbeak, *Rhyncostruthus socotranus*, that also occurs in Yemen and Oman. This bird belongs to the same group of seed-eaters as Eurasian Hawfinches and has no close relatives in Africa. The grosbeak shares with two other birds, that are wholly or nearly restricted to Warsengeli, the distinction of having been transformed by isolation. During the Ice Ages the graduated vegetation belts that surround the Arctic moved far to the south, carrying an appropriate fauna with them. Each time the ice retreated entire ecosystems crept northwards again, forcing those Eurasian animals and plants that had moved into Africa to quit their retreats. In most cases the refugees would have been unable to compete with the far more numerous and diverse indigenous forms that were reclaiming lost ground, but such recolonization from the tropics has probably been vitiated in the Horn of Africa by its forbidding desert-like hinterland. In any event, 'stranding' is the only reasonable explanation for the tenuous survival of a Eurasian Linnet and a Stock Dove on this last outpost of a continent. In the process the birds have been sufficiently altered for early taxonomists to have been baffled by their affinities. The linnet was even named after its locality, *Warsangalia johannis*; it is now regarded as an isolated species of the linnet group *Acanthis*.

Somali linnet, *Acanthis johannis*.

The Somali Pigeon's affinity with a Eurasian Stock Dove is similarly disguised. *Columba olivae* has scarlet eye masks and an iridescent copper neck. Its specific adaptation to the cliffs of northern Somalia seems to be unassailable, perhaps because ecological equivalents or potential competitors are absent. There are Rock Doves to the north, while Speckled Pigeons stop short of its range to the south and west. The Somali Little Bustard, *Eupodotis humilis*, lives in the most barren areas of northern Somali. Very closely related to Ruppell's Bustard from the Namib, this is another indication of past connections between the two areas. Heuglin's Bustard, *Neotis heuglini*, has a more extensive range in north-east Africa but also has a close relative in the southern Ludwig's Bustard.

There are many very localized populations of larks in the Horn. Larks are a rather old family of birds that have retained a generalized appearance and adaptable habits. Their success derives from tolerance of great heat, superb camouflage and the ability to survive on scarce grass, seeds and insects. All those with a predominance of grass seeds in their diet have to follow seeding seasons of the grass and are therefore nomadic. Resident populations are more insectivorous and it is among these that the endemics tend to occur. Each of the Somali larks has a different distribution pattern. The Somali Long-clawed Lark, *Heteromirafra archeri*, occupies a 200 sq km depression not far from the northern coast. This isolated relict has a relative in southern Africa. Both are probably in ecological retreat before more dominant species of lark. A Somali relative of the Sabota Lark of southern Africa, *Mirafra sabota*, has two northern sibling species, one of which, *M. gilletti*, is widespread in the Horn in rocky acacia country. Even more widespread is the local representative of the finch larks, *Eremopterix signata*. Related species occur in southern Africa, along the southern margins of the Sahara and across southern Arabia to India. Distinct desertic and woodland types occur in eastern Africa and India. In the driest north-eastern section of the Horn lives the Lesser Hoopoe Lark, *Alaemon hamertoni*. This small population probably represents a remnant of the immediate antecedents of the Greater Hoopoe Lark, *Alaemon alaudipes*, which ranges from the Atlantic shores of the Sahara to Pakistan and occurs on the Gulf of Aden foreshore where it seems to exclude the lesser species. There are three very localized Somali larks all closely related to the common and widespread Rufous-naped Lark, *Mirafra africana*. Two of these, *M. somalica* and *M. sharpei*, are confined to the vicinity of the northern massifs, the latter being of a rufous colour that matches the reddish soils common in that region. The other species is one of the two larks confined to the littoral, where powerful winds and peculiar vegetation probably force the birds to be more terrestrial. This bird, *Mirafra ashi*, is only known to occupy an area to the immediate north of Mogadishu, while the Obbia Lark, *Calandrella obbiensi*, inhabits a longer stretch of coastline.

The sandy soils of the Somali littoral may also be home to a golden mole, *Chlorotalpa tytonis*. Only known from an owl pellet picked up near Mogadishu, this animal belongs to a group of specialized and primitive insectivores that are now mainly concentrated in the Cape and south-western Africa. The golden mole's presence could signify three very different histories; one would involve relatively recent arid-axis connections with southern Africa, another possibility is an East African montane link and a third option is that Cape and Horn have had rather stable climates and soils where conservative animals can maintain themselves over millions of years. For a poor colonist the latter seems the most likely, but until the animal has been found by a zoologist rather than by an owl even its exact identity remains in question.

This coast is also home to a very small and beautiful antelope, the Silver Dik-Dik, *Madoqua piacentini*. Its habitat is the stunted bushy littoral which is about 1000 km long yet nowhere very wide, but effectively an island for some species. For them not only is the environment unique in itself but the hotter, drier hinterland is as uninhabitable as the sea that bounds its other shore. Dik-Dik evolved in the low, dry thickets of north-eastern Africa and of the four or five species, each shows progressively greater adaptation to drought. Within this spectrum the Silver Dik-Dik is the smallest and, anatomically, one of the least specialized. *Madoqua piacentini* has probably survived from an earlier era in dik-dik evolution, because the sea's cooling and moisturizing effect has reduced heat and water stress while maintaining the low-level green growth and cover these little herbivores need.

More heat-resistant antelopes occupy the Somali hinterland or Haud, among them the Dibatag or Camel Gazelle, *Ammodorcas clarkei*. This long-legged, long-

Dibitag, *Ammodorcas clarkei*.

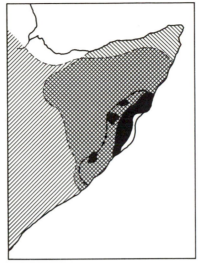

Dibitag only

Dibitag and Gerenuk

Gerenuk only

former Dibitag range

Contemporary and past distribution of Gerenuk, *Litocranius walleri* and Dibitag, *Ammodorcas clarkei* indicating decline of Dibitag into a smaller range (where, in its heartland, it is not in competition with the Gerenuk).

necked animal is one of the most distinctive of Somali endemics. It browses off the taller 'camel bushland' inland. This is a tree community called 'Ged guwah' by the Somalis and it is dominated by acacia, *A. reficens* and *A. bussei*, *Commiphora* and *Grewia*. Like the Silver Dik-Dik, the Dibatag has a more specialized and a more widely distributed relative, the Gerenuk. These relicts have a special interest for what they can tell us about the evolution of antelopes in general and their own group in particular.

The two examples just given are offshoots from small branches in the great evolutionary tree of antelopes. Much closer to the main stem is the very handsome, huge-eared antelope that is endemic to the hills and escarpments of northern Somalia. The Beira, *Dorcotragus megalotis*, is a distant relative of the rock-loving Klipspringer. It is also able to race over boulder-strewn hillsides on rubbery hooves, but its main feeding grounds are on the flats above and below these refuges. Its long limbs and slender neck are not the only anatomical suggestions of a proto-gazelle. Since the first gazelles appear in deposits dated 15 million years before the present, it can be said that the Beira offers us a glimpse into the earliest origins of these premier desert antelopes. Appropriately enough, this living fossil survives close to the geographic vortex from which the Old World's deserts radiate out. Not very much is known of their biology, but each group of animals is known to be intensely attached to its own hill or plateau. A herd of up to 10 females and young is accompanied by one or at most two males. Being so tied to home renders them very vulnerable to hunting with dogs and guns, while their need for green shoots around the year makes competition from goats and camels a still greater long-term danger. Even so, the Beira has survived because of its ability to freeze or alternatively run silently and rapidly behind the nearest ridge.

The Beira shares the lower part of its habitat with a true gazelle, *Gazella pelzelni*, which has a similar overall range but is not likely to be of any great antiquity. Instead this outpost is likely to have been colonized during a moist period by tropical gazelles of the Thomson's type, *G. thomsoni*. With increasing aridity

these retreated to an area where a marginally moister climate and richer food supply allowed a precarious existence for a declining population. The driest plateaux and bushland are inhabited by yet another Somali endemic, the much commoner Speke's Gazelle, *G. spekei*, which is allied to the Saharan Dorcas Gazelle.

This corner of Africa seems to have been a refuge and centre of evolution for desert animals, particularly antelopes. The Horn and the Cape resemble each other in being geographic culs-de-sac and in having several ecological divisions, each defined by more or less elevation, humidity, temperature and different soils. The Indian Ocean coast resembles that of south-west Africa in that marine climate, rivers and latitude combine to make discrete segments of a very long littoral. One of these segments, admittedly rather distant from the Horn, shelters yet another interesting relict antelope. The Hirola, *Beatragus hunteri*, is an early type of hartebeest. 'Early' is relative in this case, because the hartebeest family has, in evolutionary and geological time, emerged very recently. From fossils it is known that *Beatragus* were widespread and common until the true hartebeest replaced them. Today the Hirola occupies a narrow band of open grassland that is bounded by coastal forest on one side and sub-desert on the other. A waterless dry season used to keep cattle out of this area but recent provision of boreholes has brought in a fatal competitor, and cattle are now the final liquidators of a species that may have been in slow decline for nearly a million years.

Grevy's Zebra and the Wild Ass are two equines that also belong to the Somali region. The latter is really a man-made relic. Grevy's Zebra, on the other hand, has a naturally contracted range and may represent the remnant of a type of horse that once ranged across Asia from China and from northern to southern Africa. It is now almost extinct in Somalia.

The arid corridor across the tropics has been mentioned in the previous chapter. Among mammals common to north-east and south-west the Oryx, Dik-Dik and Bat-eared Fox are well-known examples. Among birds the bustards and larks have already been mentioned; the smaller hornbills and various other groups could be added. Few of these are typical of true desert and many of the most distinctive species belong to bush, thicket or thinly wooded communities. This is part of the explanation for Somalia's biological affinities being closer to the distant Kalahari desert than with the much nearer, larger but biologically poorer Sahara. The sandgrouse are an exception to this general pattern. It is true that there are endemic sandgrouse in both regions and that the Horn is the geographic centre of their range, but four of the six species that occur in the north-east also extend right across the Sahara or Sahel. This anomaly may be partly explained by the birds' survival strategy. There are grass and tree seeds, shoots and insects in even the harshest, most open desert but many of those would-be exploiters, that are not deterred by sharp-eyed hawks, snakes and carnivores, cannot cope with the heat and drought. Sandgrouse overcome these difficulties by exquisite camouflage, cautious ground-hugging movements and an ability to explode into a rocket-like flight. They are well known for their often daily flights far, high and very fast between feeding and resting grounds and distant waterholes. Some can even carry water in their plumage to dowse eggs or young away in the desert. Much of the Sahara is nearly bare of both animals and plants; here there are fewer predators and, for the sandgrouse, fewer competitors. On the desert's margins the sandgrouse still tend to use the most open degraded areas and here they can be a sensitive indicator of local impoverishment, whether natural or induced. In short, the sandgrouse is a true desert bird.

Even so, every desert animal has evolved by stages. Its emancipation from water and a more crowded community has been achieved step by step over great stretches of time. Fossil sandgrouse from nearly 40 million years ago show that

Hirola, *Beatragus hunteri*.

this process has been a prolonged one for these birds. The contemporary species that live in milder habitats could either be descended from earlier stages of sandgrouse evolution or, less likely, could have moved back into a narrower niche during the many climatic changes of the last few million years.

Of special interest in this respect is a sandgrouse that follows the course of the Rift Valley all the way from Eritrea to northern Zambia and on to the Kalahari. Very broadly the distribution of the Yellow-throated Sandgrouse, *Pterocles gutteralis*, marks out the most likely route of a desert corridor that would have linked north-east and south-west Africa. It seems more likely that this bird's attachment to a narrow string of flats and valleys is the remnant of a very ancient distribution rather than a more recent invasion from the desert. Indeed, the arid axis could have been the principal site for sandgrouse evolution. Although they are widespread in the Sahara, Arabia and southern Asia, nine of the twelve African species occur in some part of north-east or south-west Africa. That the corridor links two major desert evolutionary centres is made all the more likely because of the many other arid-adapted organisms that are paired in these two parts of Africa.

A fickle climate is typical of the whole Somali arid zone and long-term fluctuations are matched by the devastating short-term droughts that have hit world headlines in recent years. A very telling statistic, which emphasizes the role of adaptation, even in domestic animals, is the breakdown of livestock mortality in the 1974 drought. In that year the Somalis lost 40% of their camels, 60% of their sheep and goats and 80% of their cattle. This die-off allowed recovery for some over-stressed vegetation during the years that followed but the relief has been temporary. Wild vegetation and animals in this region are extremely hardy but the future of many interesting organisms will be dependent upon Somalia adopting a modern conservation policy in which the activities of people and livestock are regulated. At present this can only be a pious hope; the various peoples of Somalia are fiercely independent and quite willing to go to war. Indeed, as recently as the first two decades of this century about a third of the male population was annihilated in a civil war. For some time yet the fortunes of Somalia's flora and fauna are likely to be submerged by the tide of strictly human politics.

In spite of its turbulent history the Horn has attracted several outstanding naturalists. Between 1856 and 1868 the ornithologist Theodor van Heuglin travelled extensively through north-east Africa and subsequently wrote a book listing nearly a thousand species of birds. Thirty years later another German ornithologist, Oscar Neumann, reinforced western knowledge of the region. In the early years of this century a major naturalist was the administrator Charles Drake-Brockman, whose name became permanently linked with the region in the scientific appellation of an endemic halophytic grass, *Drakebrockmania somalensis*. Typical of his enquiring and versatile mind was the idea that the lightweight wood used by Somali nomads for their portable tent poles, *Givotia gossai*, could become a structural material for that new-fangled contraption the aeroplane. Even today, 'intermediate technology' may find many more uses for such material. Among the animals that had their first scientific description at Drake-Brockman's hands was the unique dik-dik *Madoqua piacentini*. Another administrator who became governor of British Somaliland was G. Archer, author of one of the principal works on Somali birds. Archer acknowledged his debt to two Somali colleagues, Aden Warsama and Ibrahim Sayed, whose knowledge of the area's natural history was incomparable. E. Chiovenda and J. B. Gillett provided much of the foundation for Somali botany. Of contemporaries who have helped explore and record the Horn's biological riches are the zoologist A. M. Simonetta, P. Kuchar (compiler of an up-to-date checklist of plants) and the collector C. Hemming, who was formerly an entomologist working on locust

Grevys zebra, *Equus grevyi*.

control. There are few people today who earn their living by travelling extensively and yet are bound by the nature of their job to observe very closely the intimate relationships between animals and plants. This the locust men used to do. Now vehicles, schedules, greater social distance between foreigners and locals and plain politics have all conspired to make sustained natural history in such wildernesses more detached and hurried.

A cluster of contemporary naturalists have now gathered under the aegis of the National Rangelands Development Agency where P. Kuchar, I. Deshmukh, I. Herlocker and others have worked with their Somali project leaders Mohamed Ayan and Dahir Abby Farah. A Somali Ecological Society has also been founded (which mainly functions in the south) but promotes the formation and protection of indigenous reserves and species. The Horn of Africa and its unique flora and fauna conceal vital clues to understanding the course of evolution in Africa. Somalia is a residual hot spot in a continent that has seen droughts much more extensive and sustained that those that have been experienced in recent years. The Horn of Africa is central to our understanding of numerous and ancient cycles of drought in Africa. The international community has a responsibility to assist Somalia form and carry out ambitious programmes of scientific study and conservation in this key area of Africa.

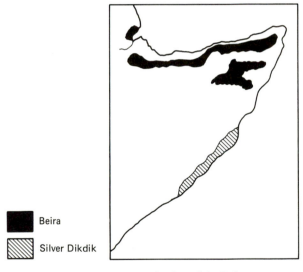

■ Beira

▨ Silver Dikdik

The distribution of the Beira, *Dorcatragus megalotis*, and the silver Dikdik, *Madoqua piacentinii*.

Opposite page. Two small antelopes endemic to the Horn of Africa. *Above,* Beira, *Dorcotragus megalotis, below,* Silver dik-dik, *Madoqua piacentini,*

Overleaf. Somali larks. *Top,* Gillet's lark, *Mirafra gilletti. Top left,* Obbia Lark, *Spizocorys obbiensis, top right,* Chestnut-headed Sparrow Lark, *Eremopterix signata. Middle left,* Archer's Lark, *Heteromirafra archeri, middle right,* Lesser Hoopoe Lark, *Alaemon hamertoni. Bottom left,* Red Somali Lark, *Mirafra sharpei, bottom right,* Ash's Lark, *Mirafra ashi.*

CHAPTER 5

Guinea Coast, Ivory Coast

Relief image of elephant on a golden Guinea.

The token was first struck in 1663 and it carried the image of an elephant. Minted in gold shipped from the coast of Guinea, it was inscribed 'in the name and for the use of Royal Adventurers of England trading with Africa'. Fifty years later it was worth 21 shillings and had become legal tender throughout the British Empire—it had long been known as the Guinea.

Thus have African commodities become fundamental units of value for Europeans and exchange of gold and ivory for manufactured goods opened four centuries of trade between West Africa and mercantile Europe. Colonization reaffirmed the Europeans' simplistic interests in Africa, two major colonies were called Gold Coast and Côte d'Ivoire. Gold mining is now nearly defunct and the tuskers too have almost gone. As we shall see, the elephants' departure may be carrying other species into oblivion with them.

Up until the early years of this century, elephants inhabited each of the ever-drier bands of vegetation that lie between the wet, densely forested coast and the dry open margins of the Sahara desert (although they flourished best in areas where they could retreat to forests or swamps for the dry season and foray out into the grasslands during the rains). Each of these vegetation belts stretches unbroken from the Atlantic to the Nile with one exception—the forest is interrupted between Togo and the Niger delta by savannah which actually reaches

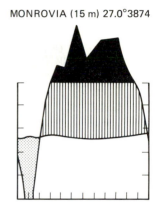

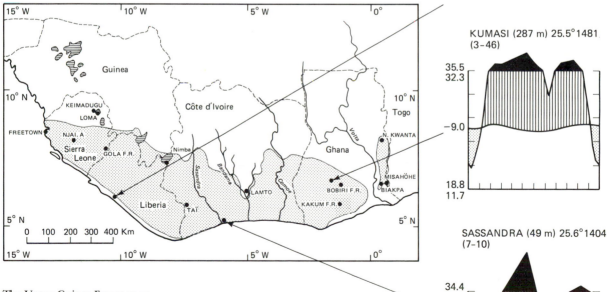

The Upper Guinea Forest zone.

the seashore. Biogeographers call the western forests the 'Upper Guinea Centre of Endemism' and the discontinuity the Dahomey Gap.

Innumerable animals and plants are common to central Africa and Upper Guinea, showing that the forests have been connected in the past. But this does not mean that the forests are identical. Even during favourable periods, the 'Gap' may have been too much of an obstacle for some species. The western forests may also have been too different from Central Africa; as a result there are prominent absentees from the west such as the Gorilla, swamp monkeys, Forest Otter (*Paraonyx congica*) and most chameleons. There are also special endemics in the west to confirm that this forest is not only isolated but intrinsically different to those further east.

The climate is one way in which they differ, being neither so consistently nor so absolutely wet. Upper Guinea is sandwiched between the hot dry Sahara and the cool dry South Atlantic, and there have been times when both arid zones compressed the coastal forests to minuscule proportions. This compression would have been most extreme during the Ice Ages, when less evaporation off the sea meant less rain and the Sahara crept south.

Even today, there are frequent reminders of the Sahara's threat. Each December the desiccating Harmattan wind blows down over the forest. If fires break out it is the north-facing slopes that burn while the moister southern slopes escape. The wind also carries a reminder of its source: sometimes desert dust falls on the red roofs of the seaside capitals of Abidjan and Accra. These oscillations between dry and wet have swung to extremes in the past. Each time entire communities would have made significant exchanges between forest and savannah. Even the African Elephant has been involved in this exchange and much more recently than is generally appreciated.

From about 3 million years ago to about 20,000 years ago the dominant elephant in the savannah was an African relative of the Indian Elephant, *Elephas recki*. The remains of this species turn up in many of the fossil deposits of the period. They replaced *Loxodonta adaurora*, a direct ancestor of the modern African Elephant, *Loxodonta africana*. *Loxodonta* was absent from the fossil record for more than 2 million years, so where were the African Elephants at this time? Almost certainly in the rain forests (where animals are seldom fossilized). During the last and most severe of the Ice Ages many very large mammals of the African savannahs died out, *Elephas recki* among them. As the climate improved the former niches of Ice Age casualties would have become available and the African Elephant was only one of several large mammals that were able to expand their range at this time. Modern survivors of the forest-dwelling elephants are the distinctive *Loxodonta africana cyclotis* which occurs, in ever-diminishing numbers, in the Liberian and Ivory Coast forests.

Other areas would have sheltered Forest Elephants but Upper Guinea was probably exceptionally well suited to them and they may have lived there for 3 million years or more. Now that they are disappearing the ecosystem can scarcely be the same without them. But does this also signify the decline or disappearance of other species? Numerous trees are known to benefit from elephants for dispersal of their seeds and these may eventually decline for lack of an efficient disperser. Others may have been favoured or discriminated against through the elephants' detailed preferences for the bark, saplings and foliage of certain species. Elephants have therefore maintained a particular mix, balance or cycle of forest trees and shrubs through their food preferences. However, a major influence on the fauna was their physical impact on the undergrowth. Browsing, pathways and 'stations' (where herds habitually congregated) all served to keep the forest floor open.

One local endemic prefers just the sort of conditions produced by elephant activity, and populations of the White-breasted Guineafowl, *Agelastes meleagrides*, are known to be in a steep decline, now exacerbated by hunting, sawmilling and human settlement. What is even more threatening for these guineafowls is the scarcity of relatively open forest floors beneath well-grown primary rain forest. In their quick, active search for food they are known to range over quite large areas. In denser undergrowth the much more abundant Crested Guineafowl, *Guttera pucherani*, takes over. This species is found right across Africa and is better adapted for forest life. It makes orchestrated group songs or

Root tangles on a rainforest floor.

White-breasted guineafowl, *Agelastes meleagrides*.

syncopated 'jam-sessions', whereas the White-breasted Guineafowl clucks and crows with a loud call somewhat reminiscent of a cockerel. This is just one of several features that link the White-breasted Guineafowl with other members of the pheasant family and it should be valued as a precious remnant of the very earliest type of guineafowl.

Its precise refuge may well be an accident of history, geography and competition from sibling species. Modern guineafowls divide along a clear savannah–forest divide and each species has adapted to the special demands of its habitat. This is probably less true for the White-breasted Guineafowl, because it derives from a period before such ecological adaptations were very advanced in guineafowls. If we are to study the details of junglefowl or guineafowl evolution or try to understand exactly how they have diverged from more distant relatives such as the curassows or guans, the survival of birds such as the White-breasted Guineafowl becomes a matter of considerable concern. Apart from contributing to our efforts to put domestic animals into a broader evolutionary context, it is even conceivable that they could provide the basis for a new domestic bird.

Upper Guinea also provides a refuge for one of the most extraordinary of forest plants, a liana which has three entirely different kinds of leaves, each with a different function, structure and physiology. *Triphyophyllum peltatum* is the very specialized relict of a primitive family Dioncophyllaceae, of which it is the only representative. It shows relationship with the south-east Asian hook lianas, Ancistrocladaceae, with the insect-digesting sundews, Droseraceae, and with the pitcher plants of Asia, Nepenthaceae. Young *Triphyophyllum* vines have

Schematic drawing of 3 stages in the growth of a *Triphyophyllum* vine. Conventional leaves on new branches sprout glandular-tipped leaves on adolescent plants in the shade, hook-tipped leaves on mature climbers.

Creepers, lianae and fungi, three life forms typical of rainforest undergrowth.

leaves that unfurl like a fern in that the midrib emerges as a coil with little or no blade. This midrib is covered with club-headed projections or hairs which, because they are sticky, collect dust, debris and insects. Exactly how these glandular leaves help the young liana remains unknown but they presumably compensate in some way for a lack of light down near the forest floor. These glandular midribs are interspersed with normal photosynthesizing leaves and also with a third type that occurs on long young shoots. In these the midrib splits at its tip and projects as two firm hooks. These assist purchase as the liana finds its way up to the canopy. *Triphyophyllum* is of special interest for what it might tell us about the early beginnings of 'insect-eating' plants.

Another aspect of the diversity and uniqueness of Upper Guinea forests is the behaviour of its trees. For example, the endemic tree *Tetraberlinia tubmaniana* readily regenerates in the shade of its own species and therefore maintains relatively pure stands in places that suit it. An exactly opposite reaction takes place in another species. Saplings of *Terminalia ivoriensis* are unable to mature anywhere near a full-grown tree of the same species. When silviculturists attempted to grow these *Terminalia* in plantations they found that the trees struggled along for 15–20 years before dying through self-poisoning.

For many tree species such periods of time are a small part of their life-cycle. As for the larger rain-forest ecosystem, many members of the community may take even longer to die, so the results of human disturbance or interference and the disappearance of elephants may take several decades or more to manifest themselves.

Terminalia is a genus with many representative species that grow outside the forest, and this interchange between forest and savannah species is not only characteristic of Upper Guinea. Throughout Africa, repeated oscillations between wet and dry have pushed animals and plants back and forth across the forest–savannah frontier, and this has affected them in various ways. Some species, such as Bushbuck, Yellow-vented Bulbuls and the Green-veined Charaxes Butterfly are able to live in a wide variety of habitats including forest. Others have developed closely related forest and non-forest sibling species. This is especially common among insectivorous bats, but there are numerous other examples: Emin's and Gambian Giant Rats, honeyguides, Pied and Crowned Hornbills, Red-bellied and Red-headed Weaver-birds are but a few.

At taxonomic levels above species most major groups of fauna are represented in both habitats and it is difficult to find any that are exclusive to forest. Even such intrinsically forest-dwelling animals as galagos, anomalures and turacos have successful non-forest forms. Immigration in the other direction is much commoner; even grasses and grass-seed predators such as finches have succeeded in invading and adapting to forest. Frequent interchanges of fauna correspond to Africa's unstable climate, but the predominance of 'derived' species in forest fauna confirms that Africa has been an exceptionally dry continent for very significant periods of time. With each return of humid conditions more recruits for the forest have come in from the thickets and dry woodlands, and the ancient residual organisms have become a small minority beside the immigrants.

As with many outlying islands, Upper Guinea shelters older relicts stranded from their parent populations to the east. From these, or from more recent invaders, Upper Guinea's own endemics may then have evolved. Among the stranded species may be those that came in during both wetter and drier climatic phases.

A very good example is the squirrels. Invading Africa less than 10 million years ago as arid-adapted ground-squirrels, they have radiated into a variety of forest forms in which the most arboreal are the most recently evolved. Upper Guinea has retained two species which probably represent earlier stages in this

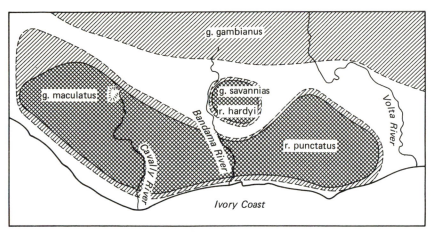

Heliosciurus gambianus, originally a savannah species.

Heliosciurus rufobrachium, originally a forest species.

Different races of two species of sun squirrel (*H. gambianus*, originally a savannah species, and *H. rufobrachium*, originally a forest species) occur together in the forests of Upper Guinea and the savannah of the Baoule Valley.

process. The Splendid Squirrel, *Epixerus ebii*, is large, soft-furred and colourful, and spends much of its time on the ground or at low levels. Its massive chisel-like teeth can open tough nuts including the local *Poga oleosa* which is almost rock-hard. The Splendid Squirrel is very rare, unlike the Giant Squirrel, *Protoxerus stangeri*, which is the largest, most arboreal of all squirrel species. It is also likely to be a very recent arrival from the east. Swamp forests are home to the smaller Slender-tailed Squirrel, *Allosciurus aubinii*, which is another endemic that may be in retreat before the invading Giant Squirrel.

Two further squirrels could be the most recent colonists. The wholly arboreal Red-legged and Gambian Sun Squirrels, *Heliosciurus rufobrachium* and *Heliosciurus gambianus*, normally split along an ecological divide: the larger Red-leg is a forest species, while the smaller Gambian is a savannah squirrel. They occur together in the Upper Guinea forests as distinct races, and in quite differ-

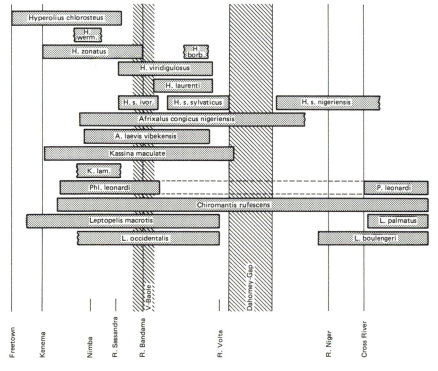

Chart of the distribution of high forest tree frogs, Rhacophoridae, in West Africa illustrating the existence of distinct sub-regions within forest zone (from Schiøtz, 1967, The Tree Frogs of West Africa).

ent forms immediately outside the forest. Although their colours are highly changeable the coexistence of two sun squirrels in the same habitat betrays the large-scale climatic changes that have taken place here. What presumably happened was that the little savannah squirrel entered the forest during a dry period, while the Red-legged Squirrel extended its range north during a wet one. Both savannah Red-leg and forest Gambian have been ecologically 'stranded', but both have managed to find a new squirrel niche and are now uniquely local forms.

One of the treefrogs seems to have made a comparable shift. *Hyperolius* are tiny, colourful frogs, an advanced and rapidly speciating group that is widespread in savannahs and along riverine forests. One of the commonest of west African treefrogs *H. fusciventris* is just such a species. In contrast, its close relative *H. wermouthi* is an inhabitant of well-shaded stagnant swamp forest; its main distinctive characteristic being the brilliant vermilion webbing on its toes and a very localized distribution within the forest proper. The latter seems to be derived from the more widespread riparian frog. Like the savannah squirrel it probably extended its range south during an arid period only to remain when a more humid climate returned.

There is also a more archaic and specialized species: Lamotte's Treefrog, *Kassina lamottei*. This is a very peculiar toad-like ground dweller that levers its broad flat body along on slender legs but, unlike other *Kassina*, cannot leap. It prefers very dense tangled forest near still water where it depends upon stealth and poisonous secretions to protect itself. The males are very rare and widely spaced out, so mating probably depends on males attracting mates with their prodigious call, which can carry for over a kilometre through the forest. Lamotte's Treefrog has become too specialized for its affinities with other

Lamottes Tree Frog, *Kassina lamottei*.

Pels Flying Squirrel, *Anomalurus peli*.

Two western hornbills. *Left top* and *bottom*, Brown-chested hornbill, *Bycanistes cylindricus*. *Right* and *middle*, Yellow-casqued Wattled Hornbill, *Ceratogymna alata*.

Kassina to be recognizable. Quite the most distinctive of Upper Guinea frogs, it is, by comparison with the *Hyperolius* species, an 'old endemic'.

The treefrogs also illustrate that the forest fauna subdivides east of the Bandama River. Even today savannah plunges deep into the forest at this point, an incursion that is known as the Baoule V. In more arid times the cut was evidently complete, providing an explanation for the separate forms of frogs, monkeys and other organisms that occur on either side of the V.

Tai Forest, Ivory Coast. Sketch of tree profiles (note the variety of form in the emergents and the undulating canopy). Some of the trees have long clean boles, others branch or support dense liana thickets.

Butterflies are mainly widely distributed rain-forest species and the insect fauna as a whole is poorer than that of Cameroon. Some species are western isolates of common types, such as the swallowtails *Papilio menestheus* and *P. hesperus horribilis*, but the Blue Banded Beauty, *Salamis cytora*, is a very distinctive Guinea endemic which deserves its English name.

The rare endemics of Upper Guinea are representative of many different levels of evolutionary history. One is the Dwarf Otter-shrew, *Micropotamogale lamottei*, which is a very primitive aquatic mammal, so specialized for life in clear forest streams that it could never have survived a drought. With another species in Ruwenzori, and neither showing any ability to colonize the 4500 km between them, this is a very clear indication of Upper Guinea having never lost its moist core. Further confirmation of the forest's age and stability comes from the presence of Black and White Anomalures, *Anomalurus peli*. These primitive gliding rodents measure 90 cm in length and weigh nearly 2 kg, which is unusually large for such a way of life. Dense, low-canopy forests are unlikely to suit this species and its main habitat is primary rain forest with tall emergents as well as palmy river-courses.

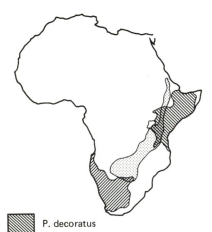

P. decoratus

P. gutteralis

P. namaqua

Opposite page, top. Somali butterflies.
Top row, Anthene janna (U&L), Acraea mirabilis (U) and Acraea miranda (L). Bottom row, Spindasis waggae (U&L) and Euchrysops migiurtiniensis (U&L).

Opposite page, bottom. Above, Blackfaced Sandgrouse, *Pterocles decoratus, bottom left,* Namaqua Sandgrouse, *P. namaqua, bottom right,* Yellow-throated Sandgrouse, *P. gutteralis.*

Overleaf. Endemic Cape butterflies (U upperside, L lowerside). *Top left, Coenyra aurantia (U), right, Aloeides caledoni (L). Second row, left Lepidochrysops trimeni (U), right, Aloeides barklyi (L). Third row, left, Capys alpheus (U), right, Poecilmitis pan (U). Bottom row, left, Phasis clavum (L), right, Phasis sardonyx (L).*

It shares this preference with the arboreal Diana Monkey, *Cercopithecus diana*, grenadier among monkeys, with a bright red, grey, black and white pattern. There is no exact equivalent for this monkey in the forests to the east. Noisy, fast, exceptionally active and very numerous in the few remaining primary forests, this monkey is a dominant member in a primate community that has fewer species than in central Africa. None the less the density of monkeys can be impressive. One of my most vivid experiences of African primates was to watch six species of monkeys (with a troop of Chimpanzees in the vicinity) feeding within sight of each other in the Tai National Park, Ivory Coast. Two colobines, a mangabey and three guenons had converged to share a single source of food; starlings, parrots, fruit pigeons, barbets and hornbills added their own whistles, shrieks, croons and honkings to the cacophony. (Two of the hornbills, the yellow-casqued *Ceratogymna elata* and the Brown-cheeked species *Bycanistes cylindricus*, are among the largest and most magnificent in a family that has nine representatives in Upper Guinea. Both are western isolates of widespread forest species.)

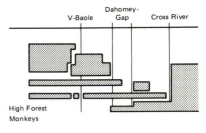

The high forest monkey fauna in West Africa showing discontinuities between species and races.

It is the monkeys that are the premier isolates in Upper Guinea. The Diana Monkey is uniquely west African, but six other monkey groups have one or more distinct forms west of the Dahomey Gap. For example, the main population of the Red Colobus, *Colobus badius badius*, ranges between Sierra Leone and western Ivory Coast with an outlying race on the Gambia River and a very different form, *C.b. waldroni* (which some regard as belonging to a different central African species, *C. pennanti*), east of the Bandama River.

A western form of Pied Colobus, *C. polycomos*, partitions into two populations, as do Collared Mangabey, *Cercocebus torquatus*, the Mona Monkey, *Cercopithecus campbelli*, and the Lesser Spot-nosed Monkey, *C. petaurista*. All these populations, including the Diana Monkey, subdivide at the Sassandra River, which is close to the point where forests narrow to less than 100 km (in the Bandama valley).

Of the colobus species the Olive, *Procolobus verus*, is a particularly archaic animal that is not closely related to the other species. The Western Pied may represent the early 'stranding' of a lineage that has continued to evolve further east. It has more conservative features than the Guercza, *C. guereza*, of central Africa. As for the Red, several phases of connection and disconnection are implied by the three populations.

A recently born Pygmy Hippopotamus, *Choeropsis liberiensis*.

The guenon monkeys will be discussed further in a later chapter but here too Upper Guinea seems to conserve earlier branches of at least three distinct radiations, each of a quite different age. The Diana Monkey is the earliest fully arboreal guenon type. *C. campbelli lowei* represents the most conservative member of the mona super-species, while *C. petaurista* may be the contemporary descendant of what I call 'the lesser peripheral monkeys', ancestors of the moustached or red-tail group.

The Olive Colobus was formerly thought to be unique to Upper Guinea, but in 1970 a relict population was found to be present east of the Niger. This reinforces the idea that Upper Guinea provides a stable refuge for species that cannot compete with the more diversified communities of the central African forests. Another animal restricted to Upper Guinea but with a tiny outlying population in Nigeria is the Pigmy Hippo, *Chaeropsis*. This 'living fossil' is another marker for stable forests in Upper Guinea. Although the hippo lies up in rivers, swamps or wallows during the day it wanders widely and I have found tracks in the Tai forests which were more than a kilometre from the nearest stream. Sometimes they follow the verges of motor tracks which become long grazing lines. The same source of low-level, fresh green growth attracts three other endemics, the minute Pigmy Antelope, *Neotragus pygmaeus*, the striped Zebra Duiker, *Cephalophus zebra*, and the much larger Jentink's Duiker, *C. jentinki*.

Jentink's Duiker is of special interest because its convergence with a typically eastern duiker suggests that Upper Guinea has been a self-contained and significant centre for the evolution of quite large forest mammals. Duikers are large-brained antelopes that are almost omnivorous although their main staple is fallen fruit. They have made an evolutionary radiation in which there is a general tendency towards ever larger size. The 'main line' duiker lineage culminates in the Yellow-backed Duiker, *C. silvicultor*, a chunky, bovine-looking animal that weighs up to 80 kg. This group of duikers is best represented in central and east-

Zebra Duiker, *Cephalophus zebra*.

ern Africa and the Yellow-back's arrival in Upper Guinea is perhaps quite recent. Jentink's Duiker is of similar size but is thought to share a common descent with the Bay Duiker, *C. dorsalis*, a predominantly western species. The evolutionary trend towards larger size would therefore seem to have proceeded independently in the main forest block and in Upper Guinea. The rarity of Jentink's Duiker implies that the Yellow-back could now be challenging its survival.

Jentinks Duiker, *Cephalophus jentinki*.

There are rather similar instances of Upper Guinea generating and/or conserving its own local forms of small carnivores. The Forest Genet, *Genetta pardina*, and Villiers' Genet, *G. villiersi*, occur here in the absence of the terrestrial Servaline Genet, *G. servalina*, of central Africa. The Liberian Mongoose *Liberiictis kuhni* is a social species thought to fill a comparable niche to the Cusimanse *Crossarchus obscurus* but is very much rarer, apparently restricted to a small part of Liberia.

Yet another local speciality is the Rufous Fishing Owl, *Scotopelia usheri*, which catches fish and amphibians in forest streams and along the margins of lakes in almost total darkness. Widespread in Africa and tropical Asia, the naked-legged fishing owls are among the least known of all birds of prey. The rufous bird shares its range (and possibly its fishing grounds) with the larger and commoner Pel's Fishing Owl *Scotopelia peli* (another small, rare species inhabits the central African forests).

Liberian Mongoose, *Liberiictis kuhni*.

Asiatic birds with minor African branches are not well represented in the far west but the cuckoo-shrikes are an exception. A complex of closely related birds occurs throughout the woodlands of Africa with the Wattled Cuckoo-shrike, *Campephaga lobata*, limited to Upper Guinea. Virtually nothing is known about this skulking inhabitant of leafy tangles in primary rain forest but it is possible that here is another instance of a woodland species being absorbed when degenerate forests made one of their periodic up-gradings. What is peculiar about this far western forest is that, although it is an impoverished community in terms of overall numbers of species and diversity of adaptive types, certain limited families or types are just as diverse as their central African counterparts. Thus, among the nine forest hornbills, eight duikers, eight squirrels and five malimbus weaver-birds, each group has one or more local endemics.

With regard to the generalities of rain-forest ecology, Upper Guinea has much in common with other areas of African rain forest, but its unique species are too numerous and too distinct to be dismissed as mere regional variants. Several species and the forest community itself emerge as more than the products of mere isolation; they are manifestations of a major evolutionary centre.

Wattled Cuckoo Shrike, *Campephaga lobata* with a pincer-like tip to the beak, specially adapted for a diet of caterpillars.

It is in the forest regions of Upper Guinea that cocoa farmers have drawn attention to the great complexity of insect ecology in rain forests. In some plantations this exotic crop is undamaged, in others ants are a major pest. Research has revealed that the forests here (and probably throughout Africa) are home to a shifting mosaic of different ant communities. One species (perhaps a tailor ant) is dominant in one patch while another genus will be the commonest in a neighbouring area. Any one patch can have up to 300 species of ants that live in a complex relationship with one another and with the animals and plants around them. Plants other than cocoa probably owe their local survival or demise to the precise balance of ants in their particular patch in the ant mosaic. Farmers have found that destroying the nests of pest species and moving in the nests of their competitors can alter the balance of power in an ant community. In this way they can promote a state that is favourable to the cocoa trees.

To study such complexities is an urgent priority but it is becoming increasingly difficult to find forests that are not in the process of being demolished. The last significant area of primary rain forest in Upper Guinea is in Ivory Coast, where the Tai National Park has been created and an important ecological study programme has begun. It is a first step towards defining the unique character of these forests. Flagrant disregard for boundaries in this park by powerful saw-millers led to two visits by Prince Bernhard of the Netherlands in 1982 and 1983. His report pointed out that the Government gave too low a priority to conservation and he called for a national conservation strategy. Since then an association for nature conservation, called 'Afrique Nature', has been founded in Abidjan under the direction of Dr Lauginie. The association produces a newsletter and organizes talks, exhibitions and slide-shows on conservation, with the ultimate aim of reaching schools and the public at large throughout Francophone Africa.

The Tai project includes a study of the forest canopy from microlight aircraft. The age, distribution, flowering and fruiting of trees is being mapped by the botanist R. Oldeman from the University of Wagenigen in the Netherlands. This university is forging links with Ivorean colleagues that will help to bring the unique worth of their forests to the attention of scholars and politicians in the Ivory Coast. It is hoped that there will be a growing awareness that the country's most precious resource has been scandalously under-valued.

The parks that are smaller than Tai are in a worse state; Marahoue National Park is under threat from commercial interests intent on new land for plantations; Mt Peko, Mt Nimba, Banco Azagny and N'Zo are all National Parks that suffer encroachments or major disturbance.

In 1983, Liberia inaugurated its first forest reserve, Sapo National Park, and there are now two more, Loffa-Mano and Cestos Senkwen. Sierra Leone is planning to up-grade a small forest reserve on its western border to national park status. In Ghana, the last major forest area, Bia Tawya National Park, has become an island surrounded by saw mills and settlement, as has the Ankasa Forest Reserve. Ghanaian conservation has the benefit of two internationally respected spokesmen, the botanist Professor E. S. Ayensu and Dr E. O. Asibey, a vice-president of the Fauna and Flora Preservation Society. Professor Ayensu has publicized the plight of these forests far beyond the borders of Ghana. In a beautifully produced and widely admired natural history guide for the young called *Jungles* he concludes: 'Lowland forests are the richest in species and are disappearing the fastest: more lowland reserves are urgently needed and must be set aside. The political and economic problems of conserving jungles must be overcome if our generation is not to be the last to see one of the most splendid and remarkable of earth's natural monuments.'

CHAPTER 6

The Oil Rivers and Bight of Biafra

River after river pours into the Bight of Biafra—called the armpit of Africa by disgruntled foreigners. It is certainly the wettest place on the continent (over 10,000 mm a year on the coast, and rain every month) and it is never less than warm. Here winds and currents conspire to pour rain from the equatorial waters of the Atlantic into the centre of Africa. Some of this rain reaches Ruwenzori,

The river Lobe emptying over falls directly into the Atlantic Ocean, South Cameroon.

more than 2000 km away, but the greatest amount falls close to the ocean whence it came. The water rushes back down the Sanaga, Cross, Mungo and a dozen other coastal rivers. Further south the great Ogooue River drains a large block of hinterland. Around Mount Cameroon, persistent rain clouds reduce the hours of sunshine, and soils that are exposed to floods and rain are very quickly leached of humus and minerals.

In spite of this the Bight of Biafra supports one of the greatest concentrations of plant and animal life on the entire continent. Of some 6000 plant species that grow in these forests many are unique to the Bight, and as many as 200 species have been recorded from a single 0.06 ha plot.

The rain forest is dominated by trees. Each tree here is the triumphant victor of a struggle with the thousands of other organisms that surround it. Each mahogany tree that grows tall enough to reach the canopy is one survivor from many tens of thousands of seeds. Its spreading crown celebrates an individual triumph, but it is the vast seed crop from which it grew that most clearly expresses the tree's genetic survival. The species' continuous evolutionary struggle is tied up in those seeds.

It is precisely this sort of botanical power-struggle that has generated the world's premier oil-producing plant, the Oil Palm, *Elaeis guineensis*. The fruits (or dates) of Oil Palms had been gathered wild for centuries, but palm oil only became a major trade commodity after the slave trade was outlawed in 1807. It was then that the innumerable creeks of the Niger delta came to be known as the Oil Rivers, and the area is still known as 'Rivers State'. By 1845 the palm nuts were being exported for soap manufacture and colonialism brought in a more systematic exploitation. By 1910 the palms that had once lined every riverbank

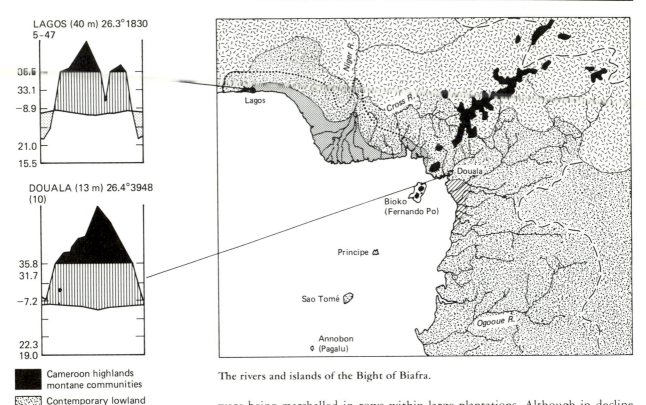

LAGOS (40 m) 26.3° 1830
5–47

DOUALA (13 m) 26.4° 3948
(10)

■ Cameroon highlands
montane communities

Contemporary lowland
rainforest zone

Swamps and mangrove
forests

Rain-forest/secondary
growth mosaic

Sand-dune forest

······· Likely extent of past
marine incursions

⌒⌒ watershed

The rivers and islands of the Bight of Biafra.

were being marshalled in rows within large plantations. Although in decline today, palm oil is still a mainstay of the region's prosperity, in spite of being overtaken by a much more lucrative oil—petroleum. Oil Rivers therefore remains an apt name.

What is not adequately appreciated, least of all by some of the local government and commercial interests, is that palm plantations, cocoa, bananas, yams and timber-milling may soon threaten the continued existence of the natural communities from which these valuable commodities were drawn. In turning the dates into margarine or soap we have detached a single product from its context, the context of a complex biological process, and we are in effect modern large-scale parasites of that process. In the case of the Oil Palm we have found a use for one member of an intensely competitive and crowded community. It is a community with a definite 'evolutionary home', a locus where conditions have been most consistent and the struggle most intense. This locus is the Bight of Biafra.

One of the winners in evolutionary struggle is the protagonist that outgrows and outlives the competition. Thus, elephants (with faster-growing, shorter-lived and smaller ancestors) have become giants of the animal world. The plants too have evolved giants, culminating in the Australian Royal Eucalyptus which can reach 114 m high. Rain-forest trees are mostly less than half that height but, like elephants, they grow slowly and last a long time. This allows the tree to accumulate scarce nutrients as they become available and thus provide the materials for a structure that can reach into the canopy and into the sunshine.

While it is Biafra's moist climate that may govern the existence of the trees, it is the trees' heavy bodies and spreading canopies that provide an environment for almost every other forest organism. What water is to fishes, and sand is to camels, so are trees to the forest community. Here more than anywhere else each species has been shaped by the crowd of other organisms in which it has evolved. The building of a single wooden column is a very great investment in time and

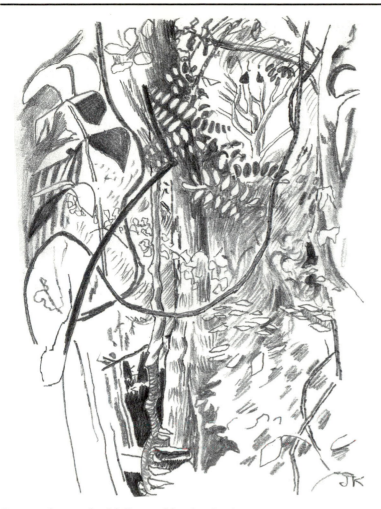

Forest undergrowth with liane and bracket fungi.

material and it is scarcely surprising that plants from many different families
find less expensive ways of reaching the sunshine. The commonest of solutions
is to become a climber, and there are areas of forest where nearly half the plant
species are lianas or vines.

Although rooted in the soil, lianas must pass through a partition before they
can flower and leaf. The partition is a dense layer of leaves; a ceiling through
which debris falls to sustain the dark, still world below. Above this is a fundamentally different environment which is exposed to the sun, the wind and a
ceaseless toing and froing of birds, insects and other animals; here there is a
rapid turnover of flowers, fruit and foliage—flux and change are normal. The
densest layer of evergreen foliage is seldom at the level of the tallest crowns
(which are often open and well spaced); instead it forms the lower boundary of a
sunny upper world. In a mixed forest, trees vary in height and in architecture,
individuals of all ages and species are well scattered so that the tracery of trunks,
branches and vines, the texture of leaves and twigs and the whereabouts of buds,
flowers and fruit are endlessly variable.

The sunlit part of the forest grows above a suspended green carpet that undu-
lates in height. A tree-fall or 'chablis' may temporarily bring the densest foliage
down to ground level in one place while the maturing of a single stand of understorey trees may lift it to more than 40 m in another; on average, however, it

Detail of a liane stem.

tends to form about half-way between the tallest treetops and the ground. By volume, each layer accounts for about half the forest, but the dynamics of these two worlds are totally different. Without sunshine plants grow slowly, they economize on their sparse rations of energy or tap it from the roots of others. For a pedestrian the undergrowth is superficially rather uniform and little of the diversity far above is apparent. In fact the diversity is there, and the insects, mushrooms and other fauna of the forest floor reflect it. The lives of most of them are strongly influenced by the unique and hidden chemistry of roots, trunks and debris.

Where the soil keeps moist, paths through the undergrowth will not go far without passing scented, leafy clumps of gingers. Indeed, the ginger and arrow-root family finds its major African Centre of Endemism in the Bight of Biafra. In Cameroon and Gabon one endemic species, *Aframomum giganteum*, has fronds that push up to 6 m tall. This is the tallest ginger plant in the world and it happens to be a favoured food of the Lowland Gorilla. The strategy for this very successful group of forest-floor plants is to survive with minimal growth until there is a break in the canopy. Because it is already rooted there, lurking in the undergrowth, the ginger plant can then rush out flowers and fruit as well as send out more rhizomes and stems during the few months that precede the leaf-layer canopy closing up again. The hot taste of ginger rhizomes probably protects the plants from attack during their quiescent periods. Not only large animals and insects but bacteria and fungi may also be deterred and it is these qualities as well as its flavour that may have commended ginger as a traditional spice and medicine.

Another swamp-loving, under-storey plant that is indigenous to the Bight of Biafra is the yam *Dioscorea*. There are several species and all protect their huge tubers with toxins which people learnt to destroy (by peeling, fermenting and cooking) many centuries ago, acquiring thereby a staple crop that enabled large numbers of people to live within the forest zone.

Cameroon is the 'centre of diversity' for some small trees that bear large edible cola-fruits. *Cola lepidota* and *C. pachycarpa* are only 2 of some 20 fruit- and nut-bearing forest trees. These not only sustain a variety of primates and squirrels but are also popular with the human population.

The larger forest trees also include a number of endemics, a high proportion of which belong to the 'cassia' or Caesalpinioideae subfamily. Their leaves are often subdivided into beautifully shaped leaflets, some very numerous and finely textured, others, like *Aphenocalyx*, simply bisected (see colour plate). All tend to have heavy pods and the endemics appear to be poor dispersers.

An advantage shared by many plants in the shade-forest is that conditions tend to remain moist, whereas the sun and the wind in the canopy quickly dry out the foliage after the regular drenchings. The sun/shade divide is therefore reinforced by a dry/moist divide, and in a rain forest there is the even greater divide of soil or no soil. Parasitic plants can take nutrients from their hosts, but parasites are scarcer than the epiphytes that find hollows to act as flowerpots or else simply spread their roots over tree-top branches. The most successful plants in this respect are orchids, which flourish in great numbers around the Bight of Biafra, because of their unique ability to cope with frequent drenching on a soil-free perch. Even so, the 'great divide' affects orchids too. Two genera, *Bulbophyllum* and *Polystachya*, flourish in sunnier, drier positions while most of the other orchids are adapted to moister, shadier sites. The latter grow long stems that are vulnerable to desiccation, whereas the former, squat and short, take advantage of each wet season to develop a pseudo-bulb crowned with two or so leaves and a flower. Each year a new shoot and flower appears and orchids of this type that have been artificially tended are known to have lived for 127 years. Theoretically the life of a single plant is infinite, but in practice the orchid

lives as long as its host branch. There is, therefore, a rapid turnover of generations and how many hundreds of species occur in the Bight of Biafra region is
still unknown. Among them are *Bulbophyllum kamerunense*, *Cyrtorchis ringens* and
Diaphananthe dorothea. *Habenaria procera* is a somewhat more widespread species
which finds its favourite perch in crevices of Oil Palm trunks. Hybridization is
at the heart of orchid horticulture, yet it seems to be very rare in the wild in spite
of very many closely related species occurring side by side. The explanation
might lie in the wide choice of insect pollinators, each of which may selectively
pollinate only one species, but the forest canopy is still too new a frontier for
this to be much more than a guess.

What is known is that most orchids require their own species of fungus to
provide sugars at the time of germination. The fungus resides in the roots of
established orchids and attacks the seed as if it were a parasite. As the orchid
develops it checks and then itself digests the fungus. The absence of their 'own'
fungus may help explain why so many orchids fail to reproduce under cultivation. The need to wait for just the right pollinator may also explain why a single
orchid can go on blooming for months (it is this, their great variety of form and
colour and their scents that have made orchids among the most sought-after and
expensive of all flowers).

Some of the nutrients that allow orchids to grow may fall from the sky as
bush-fire ashes or Harmattan dust, some may be canopy vegetable fluids but
much of it is carried up by animals. Bird and bat droppings never go to waste and
among the great variety of ants and termites there are probably several species
that build nests or transport material that orchids can get their roots into.

As in all tropical forests, ants are very abundant, but their roles in the forest
have only just begun to be explored. Special relationships with plants have been
recognized. For instance, the removal of resident ants from the Ant-plant,
Barteria, is rapidly followed by all manner of insect and fungal attacks proving
that the Ant-plant cannot survive without the ants that shelter in its hollow
stems. Other species of ants dismantle any creeper that attempts to climb over
their own host plant.

Less abundant insects, especially those in the canopy, remain largely
unknown, but the outcome of direct observation on behaviour of quite common species can have startling results. In 1977, an Oil Palm farmer in Cameroon,
noticing that Oil Palms set seed at the height of the rains, reasoned that their
aniseed-scented flowers could not be wind pollinated, as was generally assumed.
At the turn of the century an entomologist had attributed pollination to weevils, but because the weevils' interest in palms seemed wholly destructive this
was discounted in favour of the wind pollination idea. On fresh observation,
single bunches of Oil Palm flowers were found to be visited by more than 5000
weevils transferring some 3 million pollen grains in the process. The most effective of these pollinators, *Elaeidobius kamerunicus*, was identified and in 1981
exported to south-east Asia, where pollination was carried out by hand and production lagged behind Cameroon. Within a year or two palm oil production had
risen by a fifth and the extra profits now run into hundreds of millions of dollars
per year. An ironic outcome is that palm oil production in Cameroon (where
there is less sun and the soils are often depleted) is now falling behind that in
south-east Asia.

Oil Palm Weevil, *Elaeidobius kamerunicus*.

With well over 1000 species already recorded, the Bight of Biafra is richer in
butterflies than any other region. However, a very high proportion of these
range right across the Zaïre basin to Uganda, and almost as many are shared with
Upper Guinea. The forest zone between the Cross and Zaïre Rivers is the major
centre of diversity for the related genera *Charaxes* and *Euphaedra*. Both are
widely distributed and *Charaxes* are powerful fliers with relatives outside Africa.
Nonetheless it is likely that both derived from a common ancestor in Africa and

*Top row from left, Pseudacraea
acholica (upperside), Charaxes
acraeoides (upperside & underside).
Middle row from left, Euphaedera
adonina (upperside), C. fournierae
(upperside & underside), E. adonina
(underside). Bottom row, C. superbus
(upperside & underside).*

Cameroon Browed Toad, *Bufo
superciliaris*

Cameroon Caecilian, *Idiocranius
russeli.*

that the present spectrum of species includes older and younger types. It is
therefore interesting to find three closely related *Charaxes* species restricted to
this region which may be conservative relicts. Furthermore, two of them resemble other Nymphalid genera: *Charaxes fournierae* is like the local *Euphaedra
adonina*, while *Charaxes acraeoides* is strikingly similar to a *Pseudacraea*. This last
resemblance could be explained in several ways. The 18 to 20 species of
Pseudacraea include many forms that are accurate mimics of the distasteful
butterflies *Acraea, Danaus* and *Amauris*. They are very widespread and successful
butterflies, avoided by birds because of their disguise. *Charaxes acraeoides* as its
name implies, has vaguely similar markings to an *Acraea* but no 'model' within its
present range. The least satisfactory explanation is that *Charaxes acraeoides,
Pseudacraea* and *Acraea* have simply converged in their patterns. The former is
unlikely to be an eccentric maverick lineage that has embarked on a late and
feeble imitation of *Acraea*, because it belongs to the most generalized and conservative type of *Charaxes*. It seems more probable that several early Nymphalid
populations gained benefits from mimicry but that, as *Pseudacraea* refined detail
to become specialized mimics, other Nymphalids became less effective and
would have declined. If this was the case, the features that *C. acraeoides* and
Pseudacraea share may be partly due to a common Nymphalid ancestry and partly
due to both having an early history of mimicry of the same generic model.
Retention of some common ancestral features is the most likely explanation for
resemblances between *Charaxes fournierae* and *Euphaedra*. Furthermore, a tiny
enclave of *C. fournierae* in a gorge in western Uganda confirms the relict status of
this species and makes the likelihood of mimicry even more remote. A third rare
endemic may also link this early Nymphalid stock with a highly successful radiation. *Charaxes superbus* clearly belongs to the 'main line' *Charaxes* in wing shape,
tails and underwing pattern elaboration, yet all these features retain something
of the simpler, more generalized pattern of *C. acraeoides* and *C. fournierae*, sug-
gesting that these three species are close to the root of the *Charaxes* radiation.

The Bight is a centre of evolution for frogs, with eight genera largely limited
to the Cameroon–Gabon region. One of the most striking of these is the
world's largest frog, *Conraua goliath*, which is about the size of a human infant.
Two closely related species also live in this area, each graded in size and habitat.
Rocky torrents and waterfalls are preferred by the robust species *C. robusta*,
while the thick-footed *C. crassipes* lives along sluggish small rivers with sandy

Goliath Frog, *Conraua goliath*

bottoms. Here it builds itself a pyramid of sand on which it squats, eyes just breaking the surface, and waits for prey to come within striking distance. The Goliath Frog lives beside deeper waters, usually in the rocky parts of large forest rivers but sometimes beside deep pools on lesser streams. The evolution and subsequent survival of such a large frog is a token of how unique Biafra's forests and rivers are in the superabundance of water, protective vegetation and frog prey. The absence of the Goliath Frog outside the Cameroon–Gabon area not only illustrates how a frog's vulnerability is likely to increase with size, it confirms the instability of forest regions beyond the Bight of Biafra.

Some measure of the variety of birds can be appreciated from a record of 364 species in 2 sq km of forest, and the fact that the regional total approaches 500. However, the majority of species range quite widely, some to the west, others eastwards to Zaïre. One, the Dja River Warbler, *Bradypterus grandis*, lives close to the rivers and streams that shelter the Goliath Frog. The most striking and localized of the endemic birds is a bald-headed rock fowl *Picathartes oreas*, which dwells on the forest floor in areas where there are low cliffs and caves (where it nests). A close relative, *P. gymnocephalus*, has equally precise needs but is restricted to Upper Guinea.

The Woolly Bat, *Kerivoula*

Among the many spiders unique to the Cameroon–Gabon area is a colonial species that builds a mass of webs that so bind up the foliage that they become like an ornate tent which covers several tens of square metres. Within the tangled sheets of webbing and leaves other animals take refuge, notably the rare and little-known endemic bat, *Kerivoula muscilla*. While resting, the wings of these small bats resemble dried leaves and their long frizzy hair is also disguised, looking as if it were tipped with lacquer. Other bats of this genus sometimes emerge in daylight and all tend to flutter in a weak-looking flight close to the ground, hunting for what is thought to be its major food item, the tiny insects of the forest floor.

Rock Fowl, *Picathartes oreas*

Another very rare endemic mammal with cryptic nocturnal habits is *Zenkerella*, which looks like a long-legged dormouse but in fact belongs to a family that comprised the dominant rodents in Africa 30 million years ago—the anomalures. All the modern members of this family are expert gliders with

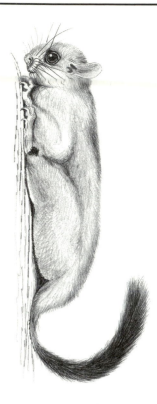

Cameroon Scaly-tail, *Zenkerella insignis*

broad membranes between their legs and very similar habits to the American 'flying squirrels'. *Zenkerella* has no trace of a membrane but possesses the same scaly tail and bat-like claws on the hind legs. The claws on the forelegs are not specialized in the same way, which suggests that this bizarre rodent rests by hanging from its widely splayed hindlegs with the scaly underside of the tail propping the body against its tree-trunk support. The animal presumably curls its head and shoulders up within this cantilevered tripod in order to rest or sleep. *Zenkerella* can probably make spreadeagled leaps between branches and it could offer a rare opportunity to study the beginnings of gliding in mammals. However, the animal is so rare and so localized in its distribution that its exact niche in this most diverse of African forests is still unknown. It is probably a canopy species.

Other anomalures use the canopy but are more widespread. The Pygmy Scaly-tail, *Idiurus*, is often called the flying mouse; spending the day in hollows of very large old trees, it emerges at night to glide from tree to tree in search of nectar and perhaps small insects. As with all canopy species very little is known about scaly-tails.

Another glider in the canopy is diurnal. It is a very beautiful black, yellow and green lizard, *Holaspis guentheri*, which has the ability to flatten its body into a disc-like shape, which it steers in flight by flexion of the body and the scaly tail. In this way it can cover distances of up to 60 m while dropping less than 20 m.

The superabundance of food and its constant renewal is one of several explanations for the large number of primates. They range from the nocturnal loris-like Angwantibo *Arctocebus calabarensis* and several species of bush-baby, through colobus monkeys, mangabeys, mandrills and guenons, to Chimpanzees and Gorillas.

Male Drills, *Mandrillus leucophaeus*, and Mandrills, *M. sphinx*, boast rainbow-tinted posteriors. In the Mandrill this is only matched by the colours of its face, colours for which students of monkey communication have begun to suggest functions and a very peculiar evolutionary history. The first point to make about these spectacular endemics is that all baboon-like monkeys must have developed as ground dwellers in relatively open habitats, and that their original

Flying Mice, *Idiurus zenkeri*

adaptation to the forest was very likely the product of becoming engulfed during periods of massive forest expansion. The combination of dense vegetation and deep shade must have made it difficult for these baboon-like animals to maintain their social system. From studies of savannah baboons we know that the numerous, heavily armed males monitor each other continuously and depend upon a simple visual signal to dampen aggression. Paradoxically this signal is a ritualized warning—the males display their weaponry by yawning at each other. This seems to release tension in the yawner and at the same time it inhibits any observer (baboon or predator) from attacking. Out in the open these yawns are easily seen and need be repeated only after a break in contact, so are relatively infrequent. In the forest, every animal is constantly losing and regaining contact and the effort that yawning uses up would become absurdly expensive of time and energy if sustained all day and every day. Part of the solution seems to have been to resort to a low-energy grimace, a sort of gargoyle's grin whereby the Mandrill draws back his lips around each canine so that they can be seen at close quarters without the trouble of a head-tossing yawn. In savannah baboons an incidental effect of tooth-baring is folding of the bare skin of the muzzle. Such details may seem unimportant, yet it is typical of animal signal systems that they elaborate and fix just such incidental visual effects, if by doing so the signal can become more efficient.

Drill

Mandrill

Total range of the Drill *Mandrillus leucophaeus* and Mandrill *Mandrillus sphinx*.

The **Agwantibo** *Arctocebus calabarensis* which, beneath the woolly exterior, is a very distinctive primate.

For the Mandrill it has been more efficient to reorganize the tissues of its face into a 'pseudo-snarl' than to waste energy in endless yawning and real snarling. Bare skin on the male muzzle has been underpadded with ridges that simulate folds; beneath them the bone has inflated to make the 'snarl' more prominent while the skin itself has become a bright electric blue so that it is more visible in the dim forest light. The muzzle has probably been elaborated in this way because upper surfaces are the best lit in poor light but the pseudo-snarl remains subsidiary to the real weaponry that it signifies. When the mouth is open, the canine teeth have a surround of red or pink lips and white whiskers but when the mouth is closed this central focus of the Mandrill's face retains a spectacular target-like marker—a perfectly circular scarlet blob over the nostrils and upper lip. In the clutter of cheeks, tufts and contrasting colours the muzzle's strongest dimension, its length, could very well become difficult to perceive. Yet, as if to counteract this disadvantage, the mature male Mandrill develops a red mid-line down the bridge of his nose. It provides a measure of the length of his enormous snout.

Understanding the full meaning of a Mandrill's face mask awaits further study in the wild and the subject will continue to intrigue analysts of animal communication. Even so, there can be little doubt that the peculiarities which so forcefully distinguish the Mandrill from baboons are linked with an originally non-forest animal having to live in dense forest. Jazzing-up colours and elaborating face structures and behaviour are part of an evolutionary solution to very special social problems associated with change in an ever-changing continent.

Mandrills forage on leaves and herbaceous growth when fruit is scarce and it is at this time that noisy aggregations of over 300 animals may gather. When fruit is abundant small, more silent groups disperse through the forest, and the overall density of animals is not high. Both species are vulnerable to disruption and fragmentation of the forest. They are also a favourite prey of local hunters, who shoot them after driving them into the trees with dogs.

The Mandrill's story is a reminder that an animal's colours, its peculiar behaviour, the detailed composition of its habitat and the climatic history of the region in which it lives are part of the inter-relatedness of life, not the isolated topics that we tend to turn them into. It is not known whether the Mandrill has been displaced by baboons in other forest areas or whether it has always been restricted to this, the richest corner of Africa's rain forest.

The Sanaga Delta

Not all Cameroon primates live in affluence and not all of Cameroon is lush and fecund. The rain that sustains so much tropical vegetation is also quick to leach soils and in at least one area there is *no* soil.

A powerful long-shore current that sweeps all the way from the Caribbean has been depositing sand in the Sanaga delta for millions of years. Over all this time the river has had to wind its way through sterile acid dunes. Changes of course have left stagnant valleys and a few ox-bow lakes. In a natural exposition of hydroponics the river's floodwaters have brought in sufficient nutrients to allow a forest to grow perched on the loose white sand-dunes. Superficially it resembles other forests; for example, there is an ebony tree, *Diospyros bipindensis*, not dissimilar to species growing in central Africa, yet the Sanaga trees have been found to have two and a half times more tannins in their leaves. In large quantities, tannins are distasteful, sometimes poisonous, and they seriously impair the digestion of any animal attempting to consume them. Most of the other trees in the delta are similarly protected, yet there is a unique species of leaf-eater, the Satanic Colobus Monkey, *Colobus satanus*, that is endemic to the Cameroon–Gabon littoral.

As the main tree-top herbivore, colobus monkeys can reach very high densities (indeed, it is tempting to describe a feeding troop as a browsing herd). In Sanaga, colobus monkeys exist at one-tenth the density of some Uganda populations because most of the forest is evergreen and inedible. They find a few ephemeral vines and rare deciduous trees but their survival has depended on a switch in diet. Satanic Colobus Monkeys have become seed-eaters. In adopting that diet they have diverged from other colobus monkeys, and the littoral forests in which they live are effectively an enclave. The Satanic Colobus specialization illustrates how the zoogeography and adaptation of animals can be influenced by the hidden properties of a plant community.

For the individual species of plant in a nutrient-starved locality there may very well be advantages that could serve them well outside their enclave and cause them to spread more widely. By intensifying the various defences that are natural to plants as they battle with each other or with the animals that eat them such species may actually speed up evolution within their own class or plant community. Sanaga may therefore be a centre of diffusion for such species.

The advantages that chemical defences confer can soon lead to the decline or disappearance of less well-protected relatives, or their restriction to situations where the defence is less useful (on cool mountains, for instance, where insects tend to be less numerous and varied). Return for an example to the ebony trees. There are nearly 40 species in the Cameroon–Gabon area, and 9 of them are found nowhere else. Among their proliferation of adaptations, differences such as those between the Sanaga ebony, *Diospyros bipindensis*, and its central African relatives are likely to be prominent. In the chemical factory that is an African rain forest, herbivores have to keep up with such defences by evolving detoxification techniques, or, like the colobus, adapt to eating a different part of the tree. Just as the defenders are likely to generate new species with each chemical improvement, so advances in the counteractions of the attackers will eventually generate new species of herbivores. For very small animals, notably those insects that live their lives out on a single food plant, the parent plant population becomes an 'island'.

Such a self-contained relationship between plant and herbivore might be capable of extension. A larger community of plants can easily become out of bounds to many herbivores because there are too few unprotected plants to sustain them. This could be an important factor in the Satanic Colobus monopoly of coastal Cameroon and Gabon. The cocktail of odd foods that this monkey can tolerate includes species that are quite specific to the area. Other species of colobus could be deterred from invasion because they lack a tolerance for local plants and do not have a taste for large numbers of seeds.

The Satanic Colobus is surrounded by other colobus populations pressing in from north, east and south. The boundaries that define their territorial 'island' can scarcely be defined geographically; they are more likely to lie in the past and in the hidden chemistry and fruiting patterns of forest trees. This example could undoubtedly be echoed by innumerable others and each would conceal a different permutation of positive and negative adaptations. Climatic and geographic influences may have determined the early evolution of many organisms, but these are now outweighed by the much more pressing influence of other animals and plants. Ultimately, it is less the climate than a direct struggle with innumerable other organisms that defines the environment for organisms in the Biafran forests, and this has special implications for their conservation.

Practical steps to conserve tropical rain forests have been very late in coming. Their study has been even slower because colonial priorities lay in the transformation and exploitation of Africa. The demand for tropical commodities gave rise to monocultures, plantations and saw mills, which were designed to replace or simply destroy indigenous ecosystems. No accommodation was envisaged.

The title of 'tropical agriculture' was bestowed on these practices rather than indigenous mixed farming. Partly because their own experience was of rudimentary man-made systems in Europe, colonial agriculturalists were preoccupied with reducing problems to manageable proportions. Active interference was always preferable to 'inactive' observation. Thus it was that the earliest attempts to study tropical forest communities came from outside the colonial establishments.

In 1962, a small field station was set up at Makokou in Gabon on the initiative of Professor P. P. Grassé of the Museum of Natural History in Paris. Today, as the Research Institute for Tropical Ecology (IRET), the station has a worldwide reputation for its work, which has found recognition in the UNESCO programme MAB and 'biosphere reserve' status for the IRET research forest. It should be remembered, however, that success was built on a pioneering association between young French field scientists and their Gabonese colleagues and guides. The association between Western scientists and IRET is maintained by workers at the universities of Brunoy and Rennes in France and it is a pity that other academic institutes, especially from Britain, Germany and America, have neglected opportunities to set up similar links with African colleagues.

In spite of such excellent and far-sighted initiatives, most efforts to study the rain forest have been left to a handful of enthusiastic botanists, naturalists and primatologists with minimal support or recognition. In the course of very long-term studies in primate ecology in Cameroon, Steven Gartlan has stimulated great interest, both locally and abroad, in plant–animal relationships. Botanists F. Mbenkum and D. Thomas and phytochemists P. Waterman and C. Mbi have developed and applied their observations, skills and knowledge of forest plants to medicine and the cultivation of new crops. In doing so they have been guided by the traditional use of plants for food and medicine.

The forests of Cameroon provide many such plants, some of them endemic to the area. 'Agusi', a thick porridge, is made of ground-up seeds from the football-sized fruits of *Treculia*. A sort of nut butter is made from the waxy seeds of *Coula edulis*, one of the large Sanaga delta trees, which are also often eaten by the Satanic Colobus. The seeds of a common liana, *Tetracarpidium*, are very popular and are sold in the markets during its fruiting season. A wide variety of cola-nuts and cola-like nuts are eaten throughout the forest zone. A small climber, *Gnetum*, provides the traditional dish 'Eru' and there are numerous other forest plants that are used by the local people.

The world at large has been initiated into the interest, excitement and charm of natural history in Cameroon and Gabon through a series of popular books. The earliest, published in the 1860s, was by the explorer Paul de Chaillu and its centrepiece was gorilla hunting. In the 1930s, Ivan T. Sanderson wrote about his

Opposite page. Some endemics from the Bight of Biafra. The canopy trees are Caesalpinaceae. *Top* (with leaf bisected into two leaflets) *Aphenocalyx margininervatus, centre* with eight leaflets, *Berlinia grandiflora,* flower, *Berlinia bracteos, left,* with dry pendant pods, *Brachystegia zenkeri. Left,* hanging, *Copaifera religiosa, centre* with up to 40 leaflets to each leaf and brown seed pod, *Monopetalanthus hadinii. Right Microberlinia bisulcata.* The leaves below the canopy belong to a wide variety of *Cola* spp. and *Sterculiacieae.* The Bight of Biafra is the major centre of endemism for this genus. Likewise many gingers, Zinziberidae, such as, from left *Megaphrynium, Costus, Afromomum, Marantochloa* and, on right, *Afromomum (longipetalum).* Endemic monkeys include the Satanic Colobus, *Colobus satanus,* the Mona Monkey, *Cercopithecus mona* and the Drill, *Mandrillus leucopheus.* Both the Forest Genet, *Genetta pardina,* and Ogilby's Duiker, *Cephalophys ogilbyi,* also occur in upper Guinea.

Overleaf. Mandrill. Illustration to show that the ritualized grimaces of the male mandrills are emphasized by a structured organization of colour and shape. Colour is also conspicuous during exaggerated postures.

Leptopelis ulugurensis

Hyperolius minutissimus

Hyperolius argus

Hyperolius parkeri

Leptopelis vermiculatus

Leptopelis parkeri

Leptopelis barbouri

Hyperolius rubrovermicularis.

Afrixalis uluguruensis

JSK.

expeditions to Cameroon to collect museum specimens, and 20 years later Gerald Durrell followed in his footsteps, but as a collector of live animals for zoos. At about this time a pioneering scientific study of rain forests was published by P. W. Richards after first-hand experience of the Nigerian forests. (*Medusandra richardsiana*, an understorey tree, found only in the Biafran forest, honours his name. This plant was discovered by a Cambridge Universiy botanical expedition to Cameroon in 1947. The trunk is fluted and, like a cocoa tree, has many shoots and white flowers which hang in clusters. It is the sole representative of a unique family—the Medusandraceae, whose name comes from its long, writhing stamens.) *Medusandra* may have an affinity with the sandalwood and African walnut families but also shows resemblances with the heather-like Cape endemic, *Grubbia*.

The first director of the IRET biological research station at Makokou, André Brosset, wrote a popular natural history of the Gabon forests in 1976, but it was with the making of two television films, 'Korup an African Rainforest' and 'Baka, people of the rainforest' that a worldwide audience of millions became acquainted with the unique value, beauty and interest of these African forests. The creator of these films, Phil Agland, and his principal scientific adviser, the ecologist Steven Gartlan, assisted the Cameroon Government in the recent formation of three Forest National Parks, Korup (between the western foothills of the Cameroon highlands and the Nigerian border), Dja and Mbam-Djerem.

The greatest concentration of forest life in Africa lies in Cameroon, so it is more than fortunate that influential citizens have seen the need for conservation and created these national parks. The leading lights in this initiative have been David Momo, Solomon Nforgwei and Nzo Ekongake.

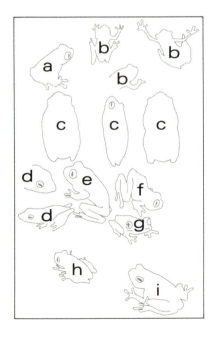

Previous page. Tree frogs from Zanj. *a*, Uluguru lichen tree frog, *Leptopelis uluguruenses*, *b*, Minute Tree Frog, *Hyperolius minutissimus*. *c*, Marbled Tree Frog, *Hyperolius argus*, *d*, Parker's Sharp-nosed Tree Frog, *Hyperolius parkeri*, *e*, Parker's Yellow-legged Tree Frog, *Leptopelis parkeri*, *f*, Vermiculated Tree Frog, *Leptopelis vermiculatus*, *g*, Barbour's Green Tree Frog, *Leptopelis barbouri*. *h*, Uluguru Sandy Tree Frog, *Afrixalus uluguruensis*, *i*, Red Spotted Tree Frog, *Hyperolius rubrovermiculataus*.

Opposite page, top. Golden-rumped Elephant Shrew, *Rhynchocyon (cirnei) chrysopygus*.

Opposite page, bottom. Fire-in-the-forest or Mango Moth, *Chrysiridia croesus*.

Details of *Medusanda richardsiana*.

Nigeria, with a larger population and more turbulent politics, has been less fortunate. In spite of pleas from local and international conservationists, the Nigerian States have permitted their precious eastern forests to be steadily eroded. The Calabar Oil Palm estates now dwarf the neighbouring Oban Forest Reserve and there has been stultifying delay in giving the important Obudu Game Reserve the proper protection and promised national park status it deserves. In a recent book on conservation in Africa the Nigerian scientist Olusegun Areola was forced to conclude that conservation does not enter into the present order of priorities for Nigeria. The Rivers Game Reserve, Stubb's Creek and Cross River Game Reserves are among more than a dozen reserves that have been degazetted in recent years. Federal and State authorities in Nigeria have therefore removed protection from the last remaining vestiges of one of the most important biological regions in Africa, the forests of south-eastern Nigeria.

The Nigerian National Forestry Research Institute has called for more protection and study of the forest remnants, but in a nation of 80 million people, with more than 1000 saw mills consuming several million cubic metres of wood each year, it will be difficult to strike a balance between maintaining an indefinitely renewable supply of timber, learning how natural forest communities function and conserving evolutionary centres which are of continental, indeed global, significance.

There are three very different perspectives: Timber, Knowledge, Conservation. Conflict between their protagonists and much confusion could be avoided if efforts were urgently directed at zoning the remaining forests according to status and function. Research and action in each of these spheres of interest could then be intensified.

No region in Africa is more productive than the Bight of Biafra, yet for biologists and politicians 'production' means very different things. The biologist observing chains of interdependence in nature finds in the Biafran forests the greatest number of interacting organisms and a complexity to their relationships that goes beyond real comprehension in the human mind, let alone the computations of a computer. The biologist is aware that he is observing processes that cover centuries. The cycle that gave rise to today's mahoganies would have already begun when the first Portuguese merchant ships sailed into the Bight of Biafra in 1471. Economists and biologists may agree in calling that mahogany 'production', but for the former it is a commodity subject to laws of supply and demand that operate in months, not centuries. For how much longer can the Bight of Biafra remain Africa's richest estate? In less than a century, high forest in Nigeria has declined from some 15% of the land to 2%. There has been

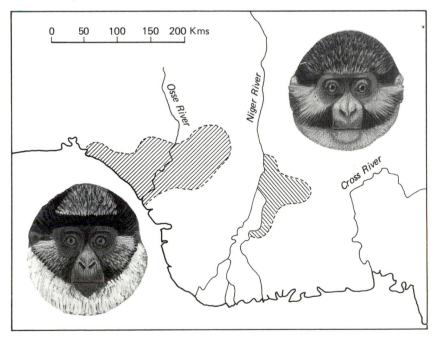

Distribution of two rare monkeys in the Niger delta region. The Red Bellied Guenon *Cercopithecus erythrogaster* to the west, and Sclater's Monkey, *Cercopithecus (cephus) sclateri* to the east.

some regeneration of secondary growth and cultivation of plantations but where will it end? The Niger delta and its hinterland has its own very localized endemics, including two unique monkeys: the Red-bellied Guenon, *Cercopithecus erythrogaster*, and Sclater's Guenon, *C. sclateri*. Their plight has focused the attention of IUCN and WWF on the remaining heavily hunted forests where they still occur.

A promising development in Nigeria has been the emergence of the Nigerian Conservation Foundation which has formed an alliance with officials of Bendel State and WWF to press for the foundation of a 70 sq km wildlife sanctuary in the heart of the White-throated Guenon's range in the Okumu Forest. The Okumu Forest Project was launched by Prince Bernhard of the Netherlands in October 1987 after intensive lobbying by Nigerian biologist Pius Anadu and primatologist John Oates. Members of the Nigerian Field Society have also supported the new venture and made visits. Following expeditions by the Conservation and Ecology Unit of University College London and Benin University, plans have been drawn up for a joint research and training programme. Against these small-scale gains must be balanced the violence of invading farmers against the Forest Department (in 1985 and earlier in 1973) and invasions by armed gangs to fell undersized timber and the arbitrary excision of 156 sq km of forest reserve to be felled and planted up by the Okomu-Udo Oil Palm Company.

Puncturing the belief that silviculturists can replace a rain forest, the Director of Nigeria's Forestry Institute, Dr P. R. Kio, has pointed out that the big trees will never again be produced, no matter what treatments are applied. He and many other Nigerians are concerned at the casual way in which a precious resource is being squandered. If southern Nigeria loses all its natural forests it will be for the first time in many millions of years. The consequences cannot be predicted.

CHAPTER 7

Zanj—Spice Coast

Ever since they were first rigged, sailing boats have travelled up and down the shores of east Africa. Between November and March for more than 2000 years the north-east monsoon winds (locally called the Kaskazi) have brought fleets of sailing ships from Arabia, India and, at one time, Indonesia. After April winds from the south-east prevail and the Kusi monsoon sweeps the little boats back to the Arabian Sea.

The ancient name for this coast is Zanj. Here timber, ivory and slaves drew the Arab and Persian fleets back year after year. Later, spices such as cloves, nut-meg and pepper also became important but from the beginning there was one commodity which was more mysterious and precious than any of these, amber-gris. Ambergris is still used for perfumery but was originally a major ingredient for fixing and enhancing exotic scents. We know now that this esoteric sub-stance comes from the jaws of squids processed in the gigantic chemical vat that is the sperm whale's gut. Amongst its many peculiarities is a very slow rate of decomposition and it also floats. There used to be hundreds of thousands of sperm whales in the Indian Ocean and each occasionally excreted gobs of amber-gris, so it can be appreciated that quite a lot of it used to bob up and down in the waves. Equatorial currents beginning off Sumatra flow all the way to Africa dumping an ocean's debris on the beaches of Zanj. Today fishermen are still enthusiastic beachcombers but the whales have been decimated and ambergris has also lost some of its mystique.

The warm currents and easterly winds that flow in towards east Africa's coast deposit more than ambergris. The rain they bring makes it one of the most favourable environments on earth. Warm, wet sea air meets the land, rises, cools and sheds moisture which in turn sustains a vegetation of extraordinary diver-sity. Over 3000 species of plants grow here, some 500 of which grow nowhere else. One reason for the botanical abundance is that this local climatic pattern is likely to have been maintained with little variation for more than 30 million years, because its source, the Indian Ocean, has been a warm tropical sea throughout this time. The east African coast has therefore had the benefit of a climatic stability that is as unusual in Africa as it is limited in extent. Inland it is dry from June to October south of the equator and wet for the intervening seven months. By contrast, the drier regions to the north only see rain during April and October. Although the equatorial section of the coast also has a dry season it enjoys some rain all the year round. Today the heaviest rain (over 2000 mm per annum) falls on Pemba Island and on higher ground inland. Behind these moisture traps there are rain-shadows, pockets of intensely arid country, and the gradients between moist forests on the hills and near-desert on the flats beyond are very steep. As in the Cape, it is this contraction of several major cli-matic zones into narrow bands that gives ecological variety to a very small area. It is the moister habitats of the Tanzanian massifs that are the primary concern here but it also needs to be remembered that these graduate, mostly in a dis-jointed mosaic of intermediate types, to arid-adapted communities on landward facing slopes. Some foothills in the rain-shadows of large mountains are near-desert they are so dry, and these have their own endemics. As was mentioned in earlier chapters, such species tend to have affinities with south-western or north-eastern Africa, but the fact that these rain-shadows may have been in existence for up to 30 million years may help to explain why some species are so localized. Soil types are known to be an important influence and the fact that dominant soils and rocks differ from one massif to another is important for all plants regardless of their water regime. For example, limestone outcrops in the Ulugurus and Uzungwas, sandstone above Lindi, while the Pugu hills are of kaolin.

On the Zanj coast, the focal centre for biological abundance is the Usambaras. It is on and around this deeply bisected massif that the greatest diversity of species is to be found. It is a diversity that is primarily due to being in quite the moistest and most favourable spot along the whole eastern seaboard. But the unique richness and diversity of the eastern Usambaras is also reinforced by isolation and a variety of altitudes over quite a large area. The next blocks of rain-trapping mountains lie not only 130 km away to the north and south but further inland. The conditions that favour the East Usambaras recur at various points along the seaboard. Northwards there are the very much lower Shimba Hills and the Sokoke and Witu Forests of the coastal plain; southwards lie the Pugu and Tundu Hills and the Rondo Plateau. However, none of these low coastal forests can compare with the Usambaras in biological diversity. Here is a choice of mini-climates or peculiar soils and a wider spectrum of plants and animals than is possible in a more homogeneous setting.

The Usambaras shelter the greatest number of species because of their most favoured position as a coastal rain-trap, but they are not unique in the area as mountains. There are other massifs in east Africa that share in having hot, moist foothills, steep temperate valleys and cooler uplands, and they share the same wet bounty blown in from the Indian Ocean; indeed, the Ulugurus which are higher are also wetter. They differ from the Usambaras in being further from the sea, in being somewhat less distant from the great mountain chains of the interior and the eastern Tanzanian mountains (often known as the eastern arc mountains) differ amongst each other in several vitally important particulars. They differ in size, altitude and age, in bedrock and in soils. Each of these dimensions of difference has its influence on the flora and fauna. The sum of

Forested valley, Usambara Mts.

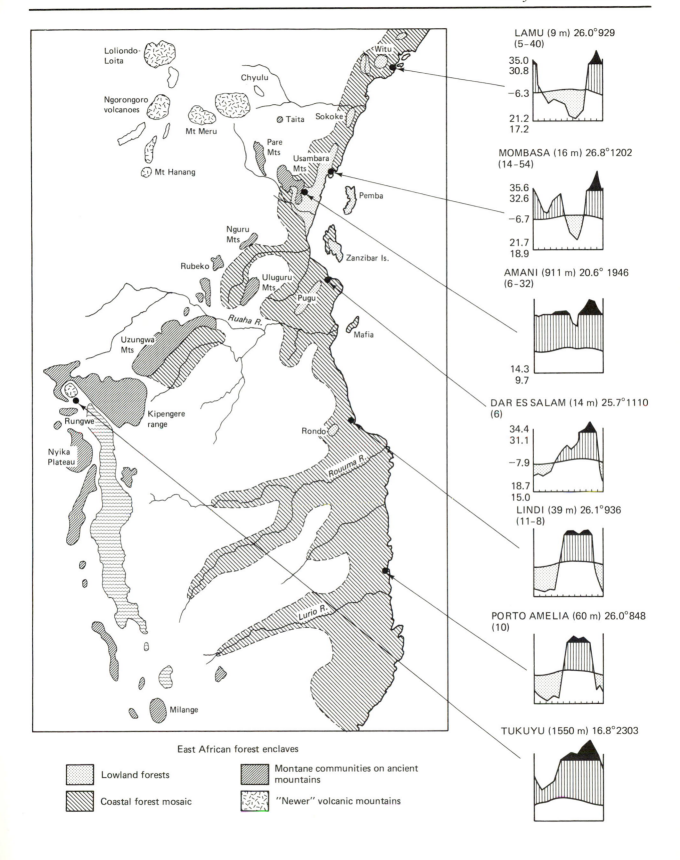

LAMU (9 m) 26.0°929
(5–40)
35.0
30.8
–6.3
21.2
17.2

MOMBASA (16 m) 26.8°1202
(14–54)
35.6
32.6
–6.7
21.7
18.9

AMANI (911 m) 20.6° 1946
(6–32)
14.3
9.7

DAR ES SALAM (14 m) 25.7°1110
(6)
34.4
31.1
–7.9
18.7
15.0

LINDI (39 m) 26.1°936
(11–8)

PORTO AMELIA (60 m) 26.0°848
(10)

TUKUYU (1550 m) 16.8°2303

Loliondo-Loita
Chyulu
Ngorongoro volcanoes
Mt Meru
Taita
Sokoke
Witu
Mt Hanang
Pare Mts
Usambara Mts
Pemba
Nguru Mts
Rubeko
Uluguru Mts
Zanzibar Is.
Pugu
Ruaha R.
Mafia
Uzungwa Mts
Rungwe
Kipengere range
Rondo
Nyika Plateau
Rouuma R.
Lurio R.
Milange

East African forest enclaves

Lowland forests

Coastal forest mosaic

Montane communities on ancient mountains

"Newer" volcanic mountains

Ternstroemia polypetala, a vestigial relict species. The map shows its disjointed distribution over central Africa.

their differences is a mosaic of communities, each with its own unique animals and plants, and it is this diversity and individuality that makes the islands, coast, hills and mountains of Tanzania some of the most exciting and important places in all of Africa. Here, each locality is like a bead of different weight and colour, and like a threaded necklace the beads are strung in separate strands. The islands of Pemba, Zanzibar and Mafia are the pendants, the coastal forests form the longest bottom strand, the old crystalline mountains further inland are the well-punctuated second row while the high volcanoes still further inland form yet another strand. Because the mountains of south-eastern Tanzania form a long line that was once called the 'African Ghats' they have been seen as the main route for colonists coming from the great forest reservoirs of central Africa. This has been true for forest birds such as the Little Greenbul and Olive Sunbird or mammals such as the Blue Duiker and Palm Civet, but these are common versatile species which, as well as being colonists, may be genetically rather stable.

Forest plants that may have spread in a similar fashion are African walnuts of the genus *Octoknema* and the aromatic *Polyceratocarpus* (although in both cases species differ from east to west, implying a long separation). The last period of connection may have been about 3 million years ago. The relationship between east coast and forest communities and those further west is not founded on one or two periods of connection or disconnection. Exchanges across degraded woodlands and thickets are also involved and the relationship within eastern Africa is also a complex story of links, frequently broken for long periods and often tenuous at best. In the Introduction it was pointed out that moisture-loving balsams, *Impatiens*, seem to have found their world centre in the very wet Ulugurus, but a great variety of distribution patterns suggests that the precise mix of adaptations in any one lineage has a profound influence on the direction, extent and success of its dispersal. In eastern Africa the patterns form a series of overlapping entities that are not exclusive to balsams. Montane communities in the Usambaras and Ulugurus have close links, as both do, to a greater or lesser extent, with the lowland coastal communities. The Usambaras have been an important source of colonists for the Plio-Pleistocene volcanoes, notably Kilimanjaro. The Ulugurus link with the Uzungwas and these communities in turn connect with those of the Tanzanian southern highlands and the Malawi mountains. The fact that these areas are discussed here, as well as in Chapter 9, is a recognition of these overlaps.

There are some very rare species that recur in both east and west African forests. For example, there is a small tree, *Ternstroemia polypetala*, which has only been recorded in Cameroon, Malawi and in two Tanzanian montane forests. It grows in a very moist atmosphere but on rapidly drained and leached ridge tops, a combination which few plants cope with. The most likely explanation for the tree's survival lies less in its own unique specialization than in the unique history of its setting. On other continents *Ternstroemia* species are still abundant but in Africa, a continent that has suffered great vicissitudes, nothing is rarer than stability itself. *Ternstroemia* grows in just such a few highly diversified localities under predictable climates. It looks as though it has squeezed into one last unassailable niche; last, because its contemporary setting could never have been widespread so that its original range must have taken in a wider choice of habitats.

Since those times it is not so much the physical as the biological environment that has changed. Innumerable new plants and many more herbivorous animals have developed and it is to competition with other trees and to plant predation or disease that we must look for an explanation of *Ternstroemia*'s rarity.

Here too will lie hidden clues to what limits many other 'old endemics', which are variously described as 'relicts' or 'specialists'. Distinguishing between the two is often neither possible nor helpful. For example, two forest hyraxes live in eastern Tanzania; one is ostensibly montane, the other a lowland species. The rabbit-like hyrax looks an unlikely climber, yet the existence of six arboreal species shows that the switch from rock to tree-dwelling has been a success. Furthermore, the species graduate in degree of adaptation to tree living. The least specialized tree hyrax, *Dendrohyrax validus*, lives among rocks and trees in the Usambara, Uluguru and Kilimanjaro mountains, while the lowland and riverine forests below are occupied by the still more arboreal *D. arboreus*. The likelihood that *D. validus* has been displaced from the lowlands by its sibling is reinforced by the fact that some trees in the lowland forests of Zanzibar and Pemba Islands are redolent with the pungent smells and weird nocturnal wails of *D. validus*. Not only is *validus* wholly arboreal in these lowland forests but *D. arboreus* becomes montane and 'returns' to rock dwelling in other parts of its range, in Kivu and Ruwenzori, where a still more advanced tree hyrax monopolises the lowland forests.

Eastern Tree Hyrax, *Dendrohyrax validus*

The important implication of this, and one that applies to a great many other organisms, is that many 'montane endemics' were once widespread species that have been left behind in those parts of their range where they can hold off competition. It is not so much cooler temperatures or special foods that cause such species to cluster around mountain massifs but rather the many sharp gradients in soils, humidity, temperature and so on that have multiplied the choice of mini-habitats.

It has recently been pointed out that many birds tend to fly up and benefit from the diversity of the mountains while the weather is favourable but return to the foothills or lowlands (whence most of them originate) when the weather gets colder. It is also significant that most highland butterflies in east Africa can be matched to a close lowland relative, as can many chameleons and mammals. Whenever there is any hint of conservative characters in one, or an advanced condition in the other, the montane form is almost invariably the more archaic type although it may also show some special adaptations.

East Africa harbours the greatest number of endemic treefrogs in the whole continent, with at least 17 species belonging to 11 genera in the Usambaras alone. Among treefrogs, one of the supposedly more archaic types, *Leptopelis*, has its main centre along the coast and in the adjacent mountains, whereas

Lindi Hawk Moth, *Rufoclanis macleeryi*

Frogfoot Moon Moth,
Antistathmoptera daltoni

Millepede species have proliferated on
the East African coasts and mountains.

Hyperolius, a very numerous group of minute frogs, is widespread and abundant throughout Africa (although it too has generated many local endemics). The latter seems to have made a breakthrough in communication. Each species has developed its own very distinctive song, and their marbled colouring is often bright and very variable. The toothed *Leptopelis* tend to be larger, to be camouflaged and to see through a vertical pupil, whereas the tiny toothless *Hyperolius* have horizontal pupils. Local species of the latter probably represent neoendemics, while *Leptopelis* are the old endemics of the treefrog world.

Those most ancient of amphibians, the worm-like caecilians, are found in the Usambaras and other massifs as well as coastal localities, where they tend to emerge from the soil after heavy rain. There are a great number of endemic reptiles and as many as 14 species of lizard are unique to the East Usambaras alone.

The insect fauna is often disproportionate in numbers to the tiny size of these forests. On one massif, the Ulugurus, 108 species of carabid beetle have been found of which 47 are endemic. A similar story can be told for many other groups. One beautifully patterned hawk moth, *Rufoclanis mccleeryi*, is only known from a single specimen found by a local medical officer in Lindi.

Hawk moths tend to be widely distributed because the majority are migrants. However, some species with a short proboscis are residents, and those that are endemic to a small area belong to this type (nonetheless McCleery's Hawk Moth is no more than the isolate of a more widespread species). A bizarre Frogfoot Emperor Moth with the tongue-twisting name of *Antistathmoptera daltoni* was once thought to be endemic to Usambara, but a related moth has recently been found in central Africa—which suggests that these moths may mark out the margins of the arid corridor where it meets the uplands. Compared to butterfly and moth faunas from central and west African forests those from Zanj are impoverished. None the less there are many endemic butterflies, among them the Blunt Emperor *Euxanthe tiberius*, the Veined White *Belenois margaritacea*, and the Policeman Skipper *Coeliades sejuncta*; many more eastern forest butterflies, perhaps the majority, range all the way to South Africa. In every instance these species have a related form in the western lowland forests and, in some cases, there is a proliferation of montane forms further inland.

The pursuit and collection of butterflies was a childhood passion of mine that began with netting of jewels such as Usambara Diadems, Mother-of-Pearls and Sapphires on seaside holidays. However, joy is an inadequate word to describe the thrill of capturing my first 'fire in the forest' moth, *Chrysiridia croesus*. Always arriving from the south, these iridescent diurnal moths would progress along the foreshore, mostly keeping above the trees and bushes, while I rushed along, net in hand, below. It is scarcely surprising that Victorians made their wings into costume jewellery, but what strange route carried this exquisite moth from the mango trees of Zanj to perch, encased under glass, on the breast of some London beauty?

The 'fire in the forest' is one of several coastal endemics that seem to have Malagasy origins. It is closely related to *C. ripheus* from Madagascar, and the absence of any African relatives is also suggestive of immigrant descent.

Ant species are numerous, especially in forests. A hint of the complexities of local ant societies has emerged from studies made in one of the most degraded of coastal habitats—a palm grove. For over a year the changing tenancy of individual palm trees by three species of ants was followed and recorded. Colonies of *Oecophylla* and *Anoplolepis* fought each other regularly with the latter tending to win where the ground at the palms' base was more open and sandy and the former winning where there was more scrub. Any one tree only harboured one species at a time. Into this battlefield *Pheidole* ants would suddenly arrive, move in and amicably share the bases of trunks with either protagonist, only to disappear again after a few months.

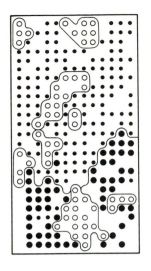
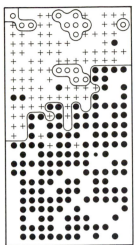
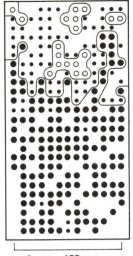

● = Palm occupied by A. longipes only

○ = Palm occupied by O. longinoda only

• Ants absent

+ P. punctulata nesting at base of palm

Approx. 130 metres

Jan 1951   Nov 1951   Feb 1952

A clear measure of great age comes from the millipedes, archaic diplopods, which are restricted to moist habitats and known to be very poor dispersers. As many as 35 species are unique to East Usambara alone and many more occur in other coastal and montane forests.

The invertebrate fauna sustains a great variety of insectivorous animals. Birds that feed on leaf litter insects are more multiform here than anywhere else in Africa. Not much is known about the details of their ecology but it is clear that the occurrence of 9 or 10 robins and thrushes in a single locality is made possible by the ecological diversity of these forests. In spite of being restricted to very small areas, the birds have an unusually wide choice of habitats and foods. There are five basic types of robin in the area. They have been given a variety of English names and their scientific names are also constantly being changed so we have to peer through a taxonomic jungle as well as a real one to perceive them. The robins demonstrate an adaptive radiation in which latitude, altitude, humidity, feeding techniques and food all play a part. They also represent some sort of evolutionary gradation in which each species pits its particular skills against sibling species which may have developed locally or come in from other forest areas, near or far. For example, Swynnerton's Robin, *Pogonocichla swynnertoni*, is a very beautiful robin with blue head and wings, orange breast and a bold crescent of black and white below the throat; it lives a tenuous existence between 850 and 1750 m on three widely spaced inland massifs. Probably the relict of a particularly archaic branch of robins, it is outnumbered throughout its range by very similarly coloured birds, the White-starred Robin, *Pogonocichla stellata*, which inhabits all the montane forests and moister woodlands in the eastern half of Africa. The White-starred Robin is probably descended from the same stock as Swynnerton's Robin and both birds have more resemblance to a Himalayan robin than to any African bird. Could they derive from a period when moist coastal thickets linked India and Africa across southern Arabia? (A sunbird, *Anthreptes pallidogaster*, and the tailor bird, *Orthotomus metopias*, are both possible shoreline colonists from the east and the tailor bird is apparently the sole representative of this typically Oriental genus in Africa.)

To return to robins, Sharpe's Robin, *Erithacus sharpei*, is the local equivalent of the Eurasian redbreast and is common in the wetter montane forests between Lake Malawi and Usambara. A drabber sibling of this species, *E. gunningi*, is found in the coastal forests, notably in the Sokoke Forest (only home to two

Competition among three species of ants (on coast below Usambara in a palm plantation). On sandy more open soils *Anoplolepis longipes* colonies dominate. *Oecophylla longinoda* colonies prefer less sandy areas of undergrowth. Colonies fight and displace one another as conditions change. *Pheidole punctulata* colonies tend to nest at base of palms and populations erupt occasionally only to disappear later. They may briefly share trees with another species (From Way 1953).

very localized birds, a scops owl, *Otus ireneae*, and Clarke's Weaver, *Ploceus golandii*).

The roll-call of robins continues with more montane endemics, notably *Cossypha anomala*, *Alethe montana* and *A. lowei*, species that show a strong preference for the drier, more marginal types of forest. Still more robins occur but they are less closely related. If we are to understand the evolution of robins as hopping, leaf-litter birds which descended, at a relatively late date, from aerial perchers like flycatchers, some clues probably lie in the biology of east African robins. What is especially interesting is the abundance of species here and their near absence in superficially similar habitats in Ethiopia (invertebrates are also thought to be much fewer in Ethiopia).

Invertebrates are extracted from the leaf litter by a variety of mammals. Ten species of shrew and four types of elephant shrew have been described from the Zanj forests. One of the most interesting of these is the Giant Elephant Shrew, *Rhyncocyon cirnei*. This elegant, long-nosed creature is unusual for an insectivore in being richly coloured and patterned. This is linked to the possession of good eyesight and a readiness to respond to visual clues and signals. On the mainland opposite Zanzibar and on the island itself is found the *petersi* form which is a rich foxy red on the forequarters with a black back and haunches. The *chrysopygus* form, from north of Mombasa, has a glistening golden rump and a red and black body (see plate on p. 124), while the much more widespread inland relations have pretty chequerboard patterns. The three are sometimes described as distinct species. They are very alert and it is difficult to creep up without being seen first. Weighing close on ½ kg, these animals find and have to consume great quantities of insects, notably ants and termites, to stay alive. Once again, their very presence is a tribute to the fecundity of the forest floor. Like a gardener digging his vegetable patch, they are sometimes accompanied by one of the local robins which are quick to dart in for the more agile insects (the robins also catch insects fleeing from driver-ant columns); meanwhile the elephant shrew gets most of its ants and termites by lapping them up with a long tongue. An archaic animal, *Rhynchocyon* is known from fossils dated at about 20 million years, in which time their basic anatomy and presumably their habits have changed very little. One secret of their success is continuous breeding. Single well-developed infants are born four or five times a year and each matures in the six to ten weeks before the arrival of the mother's next offspring.

The eastern forests are now too small to support many large mammals, although elephants, buffaloes, wild pigs and bushbuck may be abundant at times and in some specially favoured places. Two duikers are an exception to the general poverty of endemic larger mammals. Abbot's Duiker, *Cephalophus spadix*, is the eastern representative of a very widespread western form, the Yellow-backed Duiker. Original colonization of the east from further west would have been made easier by this animal's willingness to follow narrow riverine strips and to live in dry degraded types of forest. It is more than likely that it represents a relict of the immediate forebears of the larger Yellow-backed Duikers. An older relict is the small Ader's Duiker, *Cephalophus adersi*, which survives on Zanzibar Island and in the Sokoke Forest in Kenya. A general trend in the evolution of duikers has been a progressive enlargement, with the 80 kg Yellow-backed Duiker the largest and most recently evolved. Ader's Duiker, at about 18 kg, is the smallest of all the many red duikers and, like Abbot's, is likely to be the remnant of populations that were once widespread throughout the tropical African forests. Whereas Abbot's represents a very late stage in the duikers' radiation, Ader's is close to its beginnings and signifies much earlier periods of connection and disconnection with the west, as well as past linkages of Zanzibar with the mainland.

As the bird examples have suggested, not all colonization has come from the

Lesser Pouched Rat, *Beamys hindei*

Abbot's Duiker, *Cephalophus spadix*

west, nor is endemism synonymous with relict status, as is shown by one of the best-known endemic groups, the Usambara or African violet. Like many of the insects that surround them (and often serve as pollinators), these flowers are likely to be a local radiation that has responded to the peculiar stability and predictability, yet also variety, of the forest floor habitat. Of the 20 or so species of *Saintpaulia* some are local to particular massifs or isolated forests and thickets, while others are peculiar to special sites or altitudes. Their proliferation is probably relatively recent and their origins are likely to be in the east. *Saintpaulia* species belong to a family that is well represented on Madagascar and their light seeds are well suited to being carried by the wind (in fact there is one type of Malagasy *Streptocarpus* which is a strong candidate for the ancestral stock). A dramatic demonstration of how such a dispersal could have started took place on 15 April 1952. On that day a cyclone came in from the Indian Ocean and tore across southern Tanzania, felling whole forests of trees and flattening the shanty town of Lindi. Plants and animals were carried up to great heights by the storm and a ship at sea reported birds and insects raining down out of the sky on to its decks. The forests of northern Madagascar are nearly 1000 km from Lindi but the two places lie in the track of cyclones of much greater length (and the channel between the two land masses is half that distance), so to imagine Malagasy seeds flying on the wind to Africa is no fantasy. *Saintpaulias* gather their moisture at the surface rather than from underground, so that they flourish on well-drained soils in humid, heavily shaded spots. They flower all year round and are exceptionally variable, a factor which may have speeded up their radiation into many distinctively different populations. One of the rarest and much the smallest species, *S. pusilla*, has minute mauve and white flowers with equally miniature triangular leaves that are green above and purple below. By contrast, *S. grandifolia*, from the Usambara mountains, has very large corrugated leaves and abundant blue-violet blossoms. Another, *S. grotei*, has the growth pattern of a vine, sending long plant stems out in every direction for up to 1 m. In the wet season, air roots develop from nodes on these stems from which new plants can readily develop and spread.

African violets do not belong to the violet family. In common with another endemic flower, *Streptocarpus*, they are members of the family Gesneriaceae and they have an interesting history as domestic houseplants. Baron Ulrich von Paul was an aristocratic eccentric, intensely interested in his estate and gardens in Germany. His son Walter was appointed Governor of German East Africa (now Tanzania) in 1890. In order to please his father, Walter sent him some seeds of two closely related species, *S. ionantha* and *S. confusa*, from the Usambara district. From these seeds plants were grown and named in Germany, without it being immediately recognized that there were two species and many hybrids among them. The natural versatility of *Saintpaulia* was therefore unwittingly augmented by hybridization. The plants soon rewarded their breeders by throwing up new colours, shapes and sizes and, as in livestock breeding, a very human fascination in manipulating nature soon took over. However, the cultivation of these flowers remained the hobby of a small group of *cognoscenti* until the end of the Second World War. Then *Saintpaulias* really took off. They have become the horticultural equivalent of the canary or budgerigar, flowering profusely and continuously in a warm dark room to cheer the kitchens and living-rooms of innumerable people facing a long bleak winter as well as embellishing more tropical gardens further south. The number of artificially cultured varieties is now in the tens of thousands and the plants are so popular that they sustain an expanding world trade already worth 30 million dollars per annum. It is obvious that this turnover will only be maintained so long as the plant remains popular, and what is often forgotten is that all the properties that have commended the *Saintpaulia* to cultivation have been 'borrowed' and need to be artificially maintained in temperate countries. They are the natural characteristics of plants adapted to quite narrow niches in their home habitats.

In all their beauty and variety wild *Saintpaulia* are a vivid expression of quite exceptional biological interest in a few very confined and precise localities in Tanzania. Now that their climatic, geological and historical basis has been recognized, the boundaries of these highly localized communities can be drawn with some precision and the result has been considerable alarm at their imminent extinction. The world has suddenly begun to look to a small nation to take on responsibility for a series of freak enclaves where millions of years of evolution are on show. Such awareness was not developed by the colonial governments, whose conservation priorities mainly centred on big game. Huge plantations of exotic monocultures, sisal, tea, coffee, softwood and cardamom are the principal crops that have helped obliterate indigenous diversity. Plantations, to grow cash crops and the felling of timber for export, are a colonial legacy that has not died and quick returns are still a political and economic must in contracts involving foreign aid.

In 1978 at Arusha, Tanzania, Leslie Brown, the well-known agriculturalist and ornithologist, addressed a symposium on the conservation of ecological islands with the following words.

It is not only the individual farmer who casts envious eyes on the few remaining islands of original forest vegetation, it is also the large-scale development companies, government forestry and agriculture departments and international organizations. All too often such 'expert' agencies are ignorant of the ecological and the biological values of the areas they threaten.

Leslie Brown was referring more specifically to a particular project that was unique in the ignorance and audacity of its perpetrators. Consultants in the logistics and economics of timber extraction had flown out from Finland in 1976 to recommend to the Tanzanian Wood Industries Corporation (TWICO) that there were 12,180 hectares of fellable forest in the East Usambaras. The

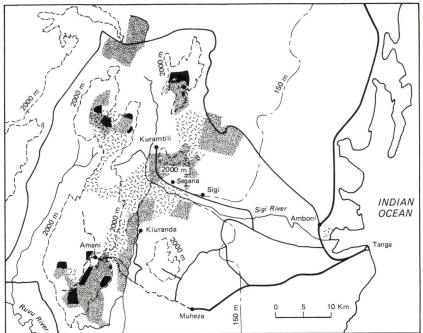

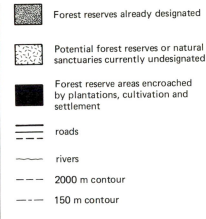

Forest reserves in the East Usambara Mts showing designated forest reserves, areas of encroachment and potential reserves or natural sanctuaries which are currently undesignated.

consultants were funded by Finland and the principal beneficiaries of the equipment, expertise and profits that would accrue were a well-established local firm of asset-strippers, already grown rich and influential from sawmilling indigenous forests. In the international furore that ensued the Finnish Government earned so much opprobrium that it eventually withdrew much of its support. A more positive outcome of this protracted scandal was that in 1983 Professor A. B. Temu, a Tanzanian botanist, and Olav Herdberg of Uppsala University set up the Usambara Rain Forest Project, and it is hoped that this and other initiatives will not only enhance the study of these unique communities but also open the eyes of young Tanzanians to the treasures they and the world are losing under the influence of 'development by plunder'.

The endemics of this region include plants that are important to local communities and could have a wider economic value. One is the Msambo tree, *Allanblackia stuhlmanni*, which has large fruits yielding fat as well as medicinal compounds. Another is a large wild nutmeg, *Cephalosphaera usambarensis*, a valuable timber tree well known for its blood-like latex.

The importance of Usambara was first recognized in 1893 by Professor A. Engler, who wrote a monumental work on the highland flora of Africa (several local endemic plants are named after him). It is on his foundations that the botanists Peter Greenway, Olav Hedberg, Tamas Pocs and Jon Lovett have all made major contributions to the knowledge of this unique floral region.

Neglect of these forests has been a sorry colonial legacy. Silviculturists in the department of forestry not only gave a low priority to conservation, but their preoccupation with large plantations of exotics tended to exclude effective community forestry. Forest reserves in Tanzania are legally open to direct exploitation and the need to change this and establish strict nature sanctuaries or 'conservation estates' has been called for by a Tanzanian biologist, Solomon ole Saibull. He believes that Tanzania cannot afford *not* to conserve, because the wilderness once lost or damaged can never be resurrected or replaced, especially by such a poor country, and this is a vision that is gaining ground among Tanzanian leaders. He has articulated these concepts on a television programme watched by millions in western Europe, in conferences and in his day-to-day work for the national parks. He has acted as a bridge between concerned audiences in the world outside and sceptical locals wanting firewood, water, meat and land wherever they are to be had.

Proposed National Parks

Natural forest

Designated forest reserve

+++ railway

— river

Forest conservation in the Uzungwa Mts, Tanzania showing remaining natural forests, proposed National Parks and designated forest reserves.

TANZANIA

Proposed Mwanihana National Park

Proposed Chita National Park

Kilombero River

rail

0 10 20 30 Km

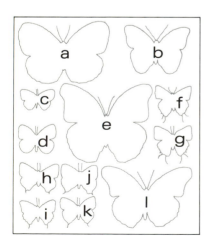

Opposite page. Zanj endemic butterflies. *a,* Dark Queen, *Euzanthe tiberius* (U), *b,* Gold Banded Euphaedra, *Euphaedra neophron* (U). *c* (U) & *d* (L), Jackson's Ringlet, *Physcaenuva jacksoni. e,* Eastern Blue Diadem, *Hypolimnas antevorta* (U). *f* & *g,* Coastal Hairstreak, *Iolaphilus maritimus. h* & *i* Crawshay's Blue, *Uranothauma crawshayi. j* & *k,* Usambara Blue, *Uranothauma usambarae. l,* Usambara diadem, *Hypolimnas usambara.*

Overleaf. The robins of Zanj. *Top left,* Iringa Robin, *Alethe lowei, top centre,* Sharpe's Robin, *Erithacus sharpei, top right,* East Coast Robin, *Erithacus gunningi. Middle Left,* Usambara Robin, *Alethe montana, middle right,* Swynnerton's Robin, *Pogonocichla swynnertoni. Bottom left* Olive-flanked Robin Chat, *Cossypha anomala, bottom right* White-starred Bush Robin *Pogonocichla stellata.*

A different example of modest bridge-building was a recent summer expedition to Tanzania from Ruthin School in Britain: they came out to make a biological inventory in the Uzungwa montane forests (a trip that fired at least one schoolboy's determination to work in tropical conservation). When such expeditions get to meet local schoolchildren, wardens and officials, the magnets for their journeys, mere plants and small animals, can take on a new significance. In the villages and classrooms of Sange or Amani a party of young students from overseas can help to make local people aware that their forests are exciting, interesting and special. There are, of course, numerous heartening initiatives; the Missouri Botanical Garden and Uppsala University are helping Tanzania build up a National Herbarium. The WWF is ready to assist in the setting up of forest national parks. A field station and small forest reserve opened recently at Mazumbai in the Usambaras. Both owe their existence to their former owners, Mr and Mrs Tanner, who ceded their estate to Dar-es-Salaam University.

The genuine concern and interest of foreign naturalists in Tanzania's unique flora and fauna needs continuous and varied expression. It needs to be public, friendly and sympathetic. Without it there will be less to deter the extinction of whole communities. Progress is being made with national-park status being considered by the Government for two remote but important forests, Mwanihana and Chita in the Uzungwas (see map), but East Usambara is still inadequately protected and undergrowth is still being cleared for cardamom crops. The map shows how small the forest reserves are in the East Usambaras and how heavily encroached. It also shows that there is still scope for consolidation and improvement. Hopes must reside in a wider recognition that the most precious bounty conferred on this land by the blue ocean beyond is not ambergris or cardamom, not red soils in the mountains nor white sands on the beaches but the ancient forest communities that cling to its seaboard and nearby mountains.

Previous page. African violets, *Saintpaulias*. *Top left, Saintpaulia rupicola, top right, S. amaniensis*; *centre, S. ionantha*; *bottom, S. grotei.*

Opposite page. An Ethiopian sketchbook. *Top left* Banded Barbet, *Lybius undatus* and Ethiopian Woodpecker, *Dendropicos abyssinicus*. *Top right* Korch Coral Tree, *Erythrina brucei* and Ethiopian Jasmine, *Jasminum stans*. *Middle left* Ethiopian Wolf, *Canis simensis*. *Middle right* Blue-winged Goose, *Cyanochen cyanoptera*. *Bottom left* Ethiopian Delphinium, *Delphinium wellbyi*, Abyssinian Rose, *Rosa abyssinica*, African Primrose, *Primula verticillata*. *Bottom right*, Giant Root Rat, *Tachyoryctes macrocephalus*.

CHAPTER 8

Ethiopia—The Fractured Dome

The Red Sea coast, a crisp clear line between desert shore and the bluest of seas, runs uninterrupted and monotonous for more than 3000 km from Suez to Massawa. It ends beneath steep escarpments that drop into an arrow-shaped bay called the Gulf of Zula. The arrow points on southwards where the escarpment continues to mark out the margin of a continent but the seashore is deflected away. Here African debris has spilled out and nearly choked the narrow sea. With each wet season more of the vast volcanic dome that is Ethiopia washes away and adds to the barren expanses of the Danakil and Awash basins. The lowest valleys are broad expanses of pale sand fed by seasonal streams lined with grey gravel. Beyond the escarpment the entire land surface is fretted into a chaos of herring-bone ridges and gullies. Above it all sit the pinnacles of Simen, a many-layered mass of lava that has been cut back to leave jagged irregular fans of ridge and trench, ridge and trench.

The Simen mountains are the last great heights; here more than 2000 m of lava is dissected out to reveal 30 million years of Stop and Go. Red bands mark the geological pauses and black the great spillages. Huge forces within the earth have piled up the mountains, but Simen's sculpture shows that the forces that are breaking it down again are equally irresistible.

Dirni gorge, Simen Mts, Ethiopia.

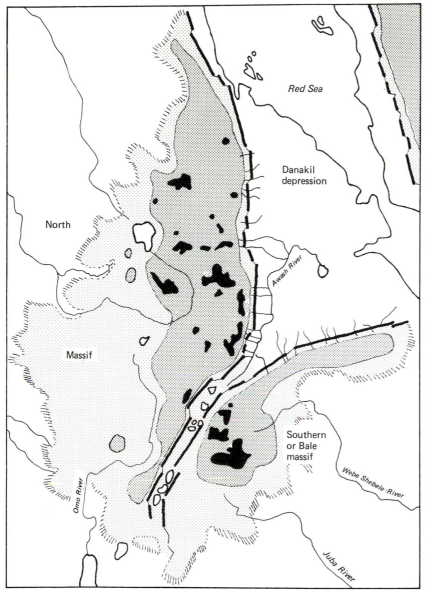

Ethiopia during the major glacial periods.

Glaciated mountains

Areas likely to have carried predominantly Afro-alpine or subalpine habitats

Areas likely to have carried montane habitats

Outer margins of the Ethiopian dome

Rivers and spill-ways off the dome and its escarpments

Rift walls

It is thought that the Ethiopian dome began to rise some 75 million years ago. The dome eventually split open and fractured three ways; each ray then slowly opened up into a valley. Today the hub lies close to the little town of Awash, one ray is the Red Sea, another the Gulf of Aden, the third the Great Rift Valley. One segment of the great dome is northern Ethiopia, the second south-western Ethiopia and the third is Yemen.

In spite of their common origin, these land masses have since followed rather different histories. That part of Africa that broke away to become the Arabian peninsula will not concern us here, but the Ethiopian massifs north and south of the Great Rift Valley have been colonized from different directions and have thus acquired some interesting and significant differences in their fauna and flora.

During the worst periods of vulcanism all this region of Africa would have been devoid of much life, but the volcanoes that had spewed out lava for so long became quiescent some 4 or 5 million years ago. Their devastations were followed by the climatic fluctuations of the Pliocene and Pleistocene, when glaciers formed on the highest mountains. During the coldest periods of the Ice Ages the glaciers would have been ringed round by broad belts of sub-alpine or chilly montane habitats; the entire Ethiopian dome would have been less like Africa and more like the tundras of Eurasia. On dry mountains the effects of cold descend further down than on moist ones, so Ethiopia has always differed in this respect from the more equatorial mountains. Ethiopia has also differed in being a much more extensive area of upland than any of the equatorial areas.

During the Ice Ages the jebels and escarpments that flank the Red Sea undoubtedly formed a tenuous link between northern Ethiopia and the Mediterranean world. It was a route travelled by many animals and plants. The southern massif, however, had a closer rift-wall connection with the Horn of Africa. Both the Ethiopian blocks face tropical Africa along their southernmost foothills, but in each case there are formidable obstacles to inhibit immigration. The White Nile and its grassy floodplains would have been an impassable barrier for many potential colonists coming from the west, especially forest-adapted species. Further east the deserts of northern Kenya would have been a comparable deterrent; and the only corridor between south-eastern Ethiopia and tropical habitats would have been along the banks of two long rivers that flow from the Bale massif to the Indian Ocean. The Omo River provided only the most tenuous of links with the south.

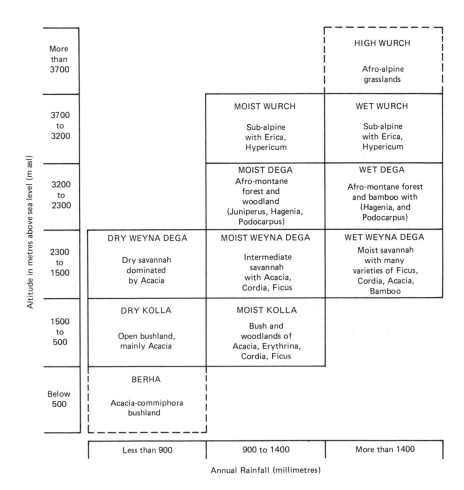

Ecological zones traditionally recognized in Ethiopia.

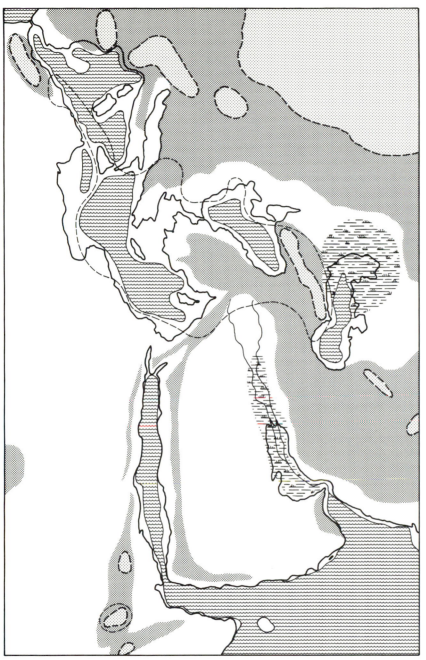

Sub-Alpine, montane and
Taiga habitats

Wetlands

Glaciated

Ethiopia's position as an outlier of the
cold habitats that dominated Eurasia
during the glacial maximum.

Botanists have been able to follow the recolonization of a volcano ever since
Krakatoa exploded in Indonesia in 1883. Sixty-one plant colonizers were recorded in the first 4 years, 276 species 40 years later. Today parts of the island
are a luxuriant forest, but its species tally is still way behind stable old islands
such as Mentawi. For all its vast size (1400 by 900 km) volcanic Ethiopia is still
relatively impoverished. Like the island of Krakatoa it cannot compare with
older localities in number or range of species and it is isolated by sea, desert and
Sudd (albeit incompletely). Unlike Krakatoa, Ethiopia has a truly unique envi-

ronment for its region, and this together with its isolation has been a potent stimulus for rapid speciation in its colonists.

Broadly speaking, the colonists seem to have had four origins. First are the local inhabitants: animals and plants of the dry lowlands surrounding Ethiopia. For more than a million years this has been the most obvious and consistent source of recruits. Then there are more distant African tropicals: inhabitants of moister lands to the south and south-west. Eurasian invaders may have flown in, crossed the straits of Bab el Mandeb, travelled along the western escarpments of the Red Sea or, least likely, come up the Nile. The fourth class of immigrants are vagrants from other cold lands that have chanced upon a habitat that closely matched their own; in shorthand jargon they were pre-adapted.

Habitats in Ethiopia are dominated by two major influences—altitude and rainfall. The Amhara people of central Ethiopia recognize the major land-types and have classified four altitudinal belts: Wurch, above about 3000 m; Dega, below it; Weyna, between 2300 and 1500 m; and Kolla, for the foothills below. These in turn are subdivided into drier or wetter types. Some of the Amharic terms correspond quite closely to botanical categories. For example, high Wurch describes the Afro-alpine steppe or meadowland that is dominated by alpine grasses, sage brush, *Artemisia*, everlasting flowers, *Helichrysum*, and the giant lobelias, e.g. *Lobelia rhynchopetalum*, which are scattered, spike-skirted, across the open uplands. Just below this frosty zone there is an infertile khaki-coloured shrubland dominated by heather, *Erica*, and St John's wort, *Hypericum*. Lower still, the woodlands and forests that once covered nearly half of Ethiopia are hard to find. They have been eroded by 2000 years of steadily expanding settlement, until they now account for a mere 2.8% of the highland area. Where the junipers still remain in settled areas, each tree stands out as an individual, Junipers have a natural variation; some droop and weep, others are frothy and ebullient, healthy trees are a deep moist green, sickly ones a dry yellowish grey. Without regeneration around them, the remaining trees also assume the character of elders brooding over their wounds. Eccentric spindly branches look like fragile bruised fingers, reaching out from the twisted gnarled solidity of old trunks while wisps of foliage hang tentatively where goats and fuel collectors cannot reach them.

Over most of Ethiopia indigenous vegetation has been annihilated. In the National Museum there is a contemporary painting by an officially approved artist that depicts the felling of a great tree to symbolize the people's triumph over feudalism and backwardness—such is the traditional enmity against trees. In at least two localities a full representation of forest types and a transition from alpine meadows to lowland scrub can still be found. It is just perceptible in the country's first National Park, Simen, where vegetation belts follow jagged courses over the ridges and gorges of this relatively small and still populated park (1500 people, 190 sq km).

The finest and most intact remnant of Ethiopia's original vegetation is contained within the proposed National Park of Bale. First mooted in 1969, the park will, if gazetted, enclose most of the high tablelands of the south-eastern massif (some 1000 sq km).

The central Sanetti plateau (about 200 sq km) is one of the most distinctive of all Ethiopian habitats. It represents the type of distinctive Afro-alpine vegetation that would have spread over a wide area of highland during the peak of the glacials (at such times Sanetti itself would have been covered in ice).

As the sizeable vestige of a once-broad tableland, Sanetti supports a few distinctive large animals such as the Great Wattled Crane, *Grus carunculatus*, the Ethiopian Wolf and its diverse rodent prey, most of which are locally evolved, plus a high-altitude hare. Although there are botanical resemblances with east Africa, Sanetti has numerous different species and they tend to form broad

Opposite. Stone pavements produced by erosion and frost in the Afro-alpine zone, Ethiopia.

- —— rivers
- —— all-weather road
- ---- local path
- principal gelada habitat
- principal ibex habitat
- ▫ National Park camps
- | National Park boundary

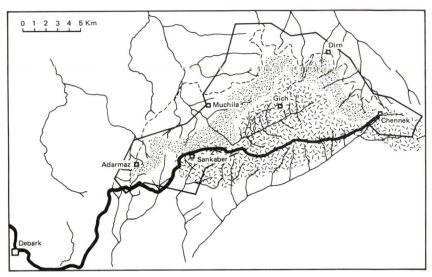

Simien water catchment protection area and National Park, showing principal gelada and ibex habitats within the park.

Haplocarpha rueppelli, a common high altitude Ethiopian endemic.

Spot-breasted Plover, *Vanellus melanocephalus*

swathes, kept short and sparse by a low rainfall, drying winds and dry season frosts that can drop to −15°C. Some striking plants are the red-hot pokers, *Kniphofia foliosa* and *K. isoltifolia*, *Lobelia minutula* and the everlasting, *Helichrysum foetidum*. Alternating frost and rapid melting cuts the thin soil into fragments and the ground heaves up each time only to collapse in the morning. Stony pavements form along imperceptible slopes where the frost and melt-waters prohibit growth. The pinnacles of Chorchora rise clean and sheer, demonstrating Ethiopia's steady erosion from the combined forces of water, wind and frost.

The great Harenna Forest which cloaks the southern slopes of the Bale massif and is only very sparsely inhabited will also be enclosed by a bold cross-section through its four graded belts. The uppermost of these is dominated by the pseudo-rhododendron *Rapanea* and tree heathers. This belt tops juniper forests in north Bale but on the moister slopes of Harenna there is instead a zone of shrubby forest which scarcely differs from that found on the upper reaches of most east African mountains. The dominant trees are *Hagenia* and the Black Cork Tree *Schefflera*. Between about 2400 and 200 m, clouds and localized rain support a dense, moist forest hung with lianes and epiphytes. Here the trees are also typical of eastern Africa, with the Osan, *Aningeria*, and African Olive, *Olea*, dominant.

The lowest, driest and much the most extensive area of Harenna typifies the sort of forest that once covered a large part of Ethiopia and possibly parts of Yemen. To this day this is prime country for honey-hunting, and it is very likely that the earliest bands of food-gatherers got to appreciate berries of the dominant understorey shrub, *Coffea arabica*, as a stimulant. Natural coffee forests were so widespread in south-western Ethiopia that an entire province, Kefa or Kaffa, bears its name. Wild coffee is still harvested extensively, but the overstorey for cultivated coffee is now more likely to be an Australian exotic than the Yellowwoods (*Podocarpus*), Stinkwood (*Pygeum*), *Croton* and Green Heart Sandalwood (*Warburgia*) that once made up the canopy of all of Ethiopia's lower forests.

Although Ethiopian habitats have much in common with the smaller mountains of tropical Africa, they are poorer in the number of species present. For example, the abundance and diversity of bulbuls, alethe robins and forest babblers, *Illadopsis*, in tropical Africa probably reflects a corresponding diversity in the plants on which the insects feed, and thus in the insects on which the birds feed. Fewer flowers and insects may help explain why there are fewer species of sunbird in Ethiopia. Likewise, blue butterflies of the Leptininae have only one species in the whole of Ethiopia, whereas there are dozens in many equatorial localities. Since this family of butterflies includes phases of dependence on specific food plants and particular species of ants during their complex and vulnerable life-history, it is obvious that their colonization will be slowed by the newness, simpleness and biological poverty of these post-volcanic, post-glacial habitats. The simpler structure of Ethiopian communities is also one of their major differences. Colder temperatures, thinner air, steeper gradients and more extensive territory combine with the removal of certain constraints (fewer predators, fewer diseases and less competition) to make Ethiopia intrinsically different from the smaller but richer highlands of tropical Africa. The first populations to adapt to major niches within such simplified habitats tend to be at an advantage thereafter; they may even deter later invaders, which may help to explain the success of several major Ethiopian endemics.

Walia Ibex, *Ibex nubianus walia*

Consider first the larger herbivores. In the 4 million years since Ethiopia became widely habitable, herbivores have been proliferating in a very late but worldwide evolutionary explosion. As one form has replaced another, peripheral refuges (such as the Cape and Somalia) have delayed or deflected the decline of older types.

Bovids have been the prime movers in this advance and one of the latest and most advanced species did succeed in invading northern Ethiopia about the time of the last glaciation. Walia Ibex are the southernmost population of the very widespread, successful and variable Nubian Ibex. In its brief sojourn in isolation from the rest of the Nubian population, the Walia has developed some minor differences which may have assumed a greater importance in the eyes of trophy hunters or Italian colonists than for biologists.

A more interesting situation is posed by the Gelada, *Theropithecus gelada*. Gelada is an Amharic name for a unique form of grass-eating monkey. Like the ibex, Gelada can use remote mountain pastures because they can clamber over cliff faces. These sociable monkeys can also withstand cold by huddling on cliffs at night, and their thick waterproof coats are of obvious utility on these wet and chilly heights. Fossils from elsewhere in Africa have shown that Geladas were widespread in Pleistocene Africa (there was even a giant form as big as a Gorilla). One reason for their decline would have been that dangerous primate, man. Geladas require long hours of undisturbed feeding to get their daily ration of grass. The Ethiopian uplands have abundant grass, few competing grazers, a reasonably tolerant human population and a ready refuge in the nearby cliffs. Troops of Gelada sometimes gather in many hundreds, and when they occur in these numbers grass leaves, stems and rhizomes are so thoroughly harvested that the temporary devastations are called 'Gelada fields'. This makes for a more dynamic and diverse community, because various fast-growing, opportunistic plants quickly invade these 'fields'.

Geladas do all their feeding seated and shuffle from one clump of grass to another without getting up. Seated feeding saves energy and conserves warmth in an animal that eats a very energy-poor diet in a cold climate. This has had extraordinary evolutionary consequences. All primates attach great importance to the genitals—this is not only the sexual interest of males in females; genital gestures (called 'presenting' in bio-jargon) appease aggression and serve as simple friendly greetings in day-to-day social life. The gesture is borrowed from sexual submission but is used by all ages and sexes, but rarely by dominant males. In the interest of saving energy Geladas have lost a primary mode of keeping the peace and also made it more difficult to monitor females' oestrus. The latter problem has been solved in Geladas by the females developing an area of bare

skin around the breasts and up to the neck which becomes distended with liquid at the same time as similar puffy, pink vesicles around the vulva. This chest ornament is a very passable imitation of the latter and both sets of vesicles are triggered by sexual hormones and thus serve as a highly visual advertisement of oestrus. The social function, as distinct from the sexual, of this visual sign is made obvious by the fact that males also develop a bare pink bosom. It is such a striking feature of Geladas that they have been called 'bleeding-heart' monkeys. Of course males cannot be mistaken for females; for one thing they are twice the size. The male's bleeding heart is simply a matter of possessing the means to attract as well as to repel other monkeys of both sexes. The ability to repel is obvious enough; every mature male has a massive cape and mane that can double his size. Combined with toothy snarls, yells, charges and bouncing off rocks and bushes, a male can put on an imposing show in defence of his harem of up to six females and their young. Yet Geladas are among the tamest and most peaceable of monkeys. In fact, it is the marginal nutritive value of their diet and the long time needed to gather it that demands an orderly society. Because the grass is everywhere, there are no quarrels over food or space. The only obvious competition is amongst males, and that is part of the regular turnover of males maturing and declining, acquiring and losing harems.

The imposing appearance of males probably has more subtle uses in the day-to-day maintenance of order. On the principle of 'armed truce', males indulge in spectacular bluffing displays but avoid fighting because of the ever-present danger that even quite young males can inflict serious damage if forced to bite. Males are reminded of this possibility in the form of a unique display, the Gelada lip-flip. It is made by any male but most particularly by those that find themselves on the defensive. It consists of exposing sharp canines and pink gums through the sudden flipping back of an exceptionally flexible upper lip. The leverage for this display comes from muscles on the muzzle. These give even the most impassive Gelada face an expression that is, to human eyes, strained and drawn.

Like so much else in Gelada biology such appearances are deceptive and conceal a wealth of hidden meanings. The Gelada is not just a bizarre Ethiopian endemic. Their densely aggregated harems demonstrate principles of social order; their feeding habits and day-to-day regime illustrate ecological economy; while their anatomy and behaviour combine to show the extraordinary ingenuity of evolution. Today Geladas are restricted to a few highlands in the northern massif; the southern highlands have a very different herbivore, the magnificent Mountain Nyala, *Tragelaphus buxtoni*.

As with the Gelada, there is fossil evidence from elsewhere in Africa to suggest that the Nyala survives in slightly altered form from an animal that was widespread 2 or 3 million years ago. While vulcanism and glaciers made the mountains uninhabitable, fringes of lower land may have sheltered Nyala and Gelada long after they had declined elsewhere. The reason for their survival may lie less in the peculiarity of the mountains than in the ring of rivers, swamps and deserts that insulate Ethiopia from tropical Africa. Neither the Eland, which is known to be recently evolved, nor the Bongo has crossed these barriers. Mountain Nyala resemble both these antelopes in feeding habits and ecology, indeed the upper slopes of Mts Kenya and Kilimanjaro, which most resemble Bale, provide a prime habitat for Eland, while the equivalent uplands on the Aberdare Mountains in Kenya are frequently visited by Bongo from the forests below. Today the Mountain Nyala ranges over the full spectrum of habitats in the Bale massif but is most concentrated in a single place, the Gaysay Valley. Outside this enclave these antelopes seem to be vulnerable to disturbance and predators, and are therefore much rarer there.

Cold and uninviting, the Bale mountains had few permanent settlers before

Opposite. Geladas, *Theropithecus gelada*

Two bull Mountain Nyala, *Tragelaphus buxtoni* in ritualized dominance display.

the 1960s. At that time a Belgian sheep farmer moved in to exploit the rich pastures and permanent water of the Gaysay Valley. The Nyala, for a long time in decline from excessive hunting, were now forced out of their last stronghold and the ranch headquarters at Dinsho soon became a market town on the new access road which was the well-trodden pilgrims' route for a local cult. By 1971 a resident observer felt fortunate to get distant glimpses of a dozen or so animals in a month. Rigorous protection and an end to sheep farming has changed that and allowed the Nyala to recover their numbers. Dinsho village is still a frogs' pond of noise, whining lorries, yelling children, barking dogs and that ubiquitous sound of the Ethiopian highlands, axes chopping wood; yet the presence of an efficient team of game wardens has transformed the surrounding valley. In Gaysay a morning trek may take the rider past several hundred Nyala.

A novice could be forgiven for thinking that the animals suffer from terminal influenza; their alarm call sends a spray of droplets from the nose and emerges as a ripe cough clearing the throat of phlegm (in fact the animals are perfectly healthy). Males are heavily built with thick necks and heavy horns that describe a lazy spiral. The white crest running from nape to tail is normally limp and barely visible, but when two evenly matched bulls meet, crests are raised and a wary circling commences. Head lowered, fluffed tail raised, each bull struts, shuffles, feints until one or the other gives way or, more rarely, takes on the challenge with a savage clash of horns. The hornless females are much lighter with long slender necks and enormous ever-switching ears behind the alert black eyes. Like the males, their steps are slow and measured but more tentative and loaded with nervous, springy energy.

The upland pastures are full of aromatic herbs, and Nyala feed on many of them including Lady's Mantle, *Alchemilla*, a *Geranium*, the powerfully scented Catmint, *Nepeta biloba*, and a feathery sage brush, *Artemisia afra*. Their indifference to the protective compounds that deter other herbivores suggests a population of animals that has become well adapted to an extreme environment, but their catholic diet, use of other habitats below the mountain-tops, and their lack of any obvious specializations tend to confirm that this is more of a relict population than a specifically montane specialist.

Highland Ethiopia provides an interesting contrast with lowland Cameroon as a Centre of Endemism. In the latter, an all-pervasive biological environment shapes a species' niche and its evolution more than the climate. In Ethiopia, diets and levels of competition or predation may be very different from lowland habitats, but it is the demands of an extreme climate and terrain that influence speciation more than the impoverished fauna and flora around it. This is particularly true for the smaller organisms—large mammals are mobile and more able to insulate themselves against the elements. For example, the Nyala only alter their behaviour in quite minor ways to avoid excessive cold or overheating. They avoid frost by lying up in heather thickets or beneath the dense skirts of juniper trees. On sunny days they are under shade by midday, often returning to depressions in the leaf litter that have been hollowed out by repeated use. If contemporary populations of Nyala and Gelada have changed their behaviour more than their basic anatomy or physiology, the same cannot be said of the smaller organisms.

For rodents, high-altitude environments are so demanding that most species are subject to rapid selection. Most high mountain areas in Africa tend to have some speciation or sub-speciation in rodents, but nowhere is this more true than in Ethiopia. In the case of the grass-rat *Arvicanthis blicki*, the genus is known to be of very recent origin so that speciation within such a new rodent type suggests that adaptation in a fast-breeding, predator-prone animal may be exceptionally rapid.

This grass-rat is part of a rodent community known to number 14 species, 8 of which are endemic to upland Ethiopia. Grass-rats and root-rats can reach astonishing densities in Afro-alpine steppes and one large endemic, the Giant Root-rat, *Tachyoryctes macrocephalus*, has been estimated to number 2600 animals per sq km on the Sanetti plateau in central Bale. In contradiction to their name, these soft-furred, golden-hued rodents gather plant parts from *above* ground, digging numerous many-branched foraging tunnels to help minimize the distance and time each animal stays on the surface. Giant Root-rats are very cautious, and browed head-top eyes survey the surroundings from a burrow's mouth before the rat emerges. They are not entirely defenceless: a fast backward dash or a bone-cutting bite from their yellow chisel-like incisors will put off any predator that is too slow or too weak to disable a muscular animal that weighs nearly a kilogram.

It is quite likely that Honey Badgers occasionally dig out Giant Root-rats from their burrows at night, but their main enemies are more likely to be above ground and diurnal, since they gather their food by day. Migrating eagles might take a few, but their main predator comes from an unexpected quarter; it is the Ethiopian Wolf, *Canis simensis* (more commonly misnamed the Simien Fox because of its red colour).

At first glance it is odd to find a courser (running is a primary characteristic of any canid) earning its living off subterranean rodents. Closer observation reveals why a canid is peculiarly well designed to live off rats in high mountain steppes. For a start, all surface dwellers are very exposed. They must be sufficiently well camouflaged or large enough to escape eagles and they must be well insulated against the cold. Unlike the smaller, sandy-coloured jackals that haunt the scrub and woodlands, Ethiopian Wolves are conspicuous. They make no effort to conceal themselves and their red and white figures stand out in the grey, khaki or tawny landscapes of the Bale and Simen. They even rest curled up in open hollows. Although the wolves' prey demands solitary hunting, pairs often travel within sight of each other, each alert to the other and to the ubiquitous rodents all around. Because an actively feeding rat soon resurfaces after a scare, the wolves' main strategy is to observe the bolt-hole and then wait motionless beside it. A wolf is not easily deflected from its ambush, and one that I walked up to allowed me to approach within a few metres before it turned its head. Looking me straight in the eyes, it startled me with a very loud ululating scream and then trotted away.

The wolf is well equipped to catch rats. Unlike the jackal, which has a conical, tapered muzzle, the Ethiopian Wolf's jaws are fronted by a flat, trap-like line of pointed incisors and long splayed canines. Behind are sharp, well-spaced teeth in an exceptionally long tooth-row. It is an arrangement that is designed to inflict a deeply damaging bite. It could have developed to cope with the special problem of despatching relatively large biting rodents but is equally possible that the mechanism has been modified from a more general use as a clamp on struggling live prey. Could the wolf and its prey have been larger in the past? Hints of such an ancestry appear in occasional attempts at the pursuit of young antelopes and hares by more than one wolf and by reports of a lamb's back being broken by a single wrench of the wolf's jaws. However, observations of such joint efforts and of kills are very rare, so we must conclude that the modern Ethiopian Wolf is pre-eminently a rat-catcher, no matter what his ancestors did.

The Giant Root-rat belongs to a family that is of Eurasian origin and has only succeeded in colonizing the cold highlands of eastern and central Africa. Other Ethiopian endemics that originated in Eurasia are the Highland Rose, *Rosa abyssinica* (which is distinguished by sweet-scented foliage as well as flowers), and *Primula verticillata*, the only African primrose, restricted to highland valleys where it grows under shade on mossy seepages. The butterfly *Pieris brassicoides* is another example of a single African representative of a Palaearctic genus. The glaciated past of Ethiopia's mountain-tops and their closeness to Eurasia are evident in the Palaearctic plankton of the highland lakes. Endemic butterflies in Ethiopia tend to be highland isolates of widely distributed African types, notably the wanderers, *Acraea*, swallowtails, *Papilio*, and *Charaxes* (see plate on p. 177). One *Charaxes* that is normally a lowland, open country species has adapted to montane forest in Ethiopia, the only area in which it occurs in such a habitat.

A butterfly that comes from the east is the Indian *Argrus hyperbius*, but this is essentially a contemporary vagrant.

A chance landfall from much further afield is the only explanation for one of Ethiopia's most interesting endemics, the Blue-winged Sheldgoose, *Cyanochen cyanoptera*. Sheldgeese evolved in the alpine and temperate grasslands of South America, where they were once a dominant grazer numbering many thousands, perhaps millions. During the Ice Ages these birds would have been very abundant and ranged much more widely; furthermore, vagrants would have survived longer and so could have wandered more widely. Even today shelducks and geese are great wanderers. For example, an immature European Shelduck was seen near the Cape of Good Hope on 23 June 1974. If the bird followed the coastline it might have travelled more than 11,000 miles in a matter of two months or less. All sheldgeese go through a month or so of flightlessness during the moult, so vagrants are soon eliminated in inappropriate environments. In the highest reaches of Ethiopia sheldgeese would have encountered relatively safe and empty uplands where the meadows, riverbanks and rarified air were similar to the Andes. The Ethiopian birds have evolved into a distinct species but still retain much in common with their South American cousins, including bizarre courtship postures. With his blue wing-patches prominently exposed the male struts before his mate, neck arched back as far as possible while his beak points to the sky, emitting breathless whistles. Thick, loose plumage retains warmth and makes the bird assume various amorphous globular shapes. The birds often prefer to flee on foot rather than on the wing. When they do this they sometimes remind me of elderly shoppers laden with bags and hastening to catch a departing bus. Their composure may be the result of practice during the moult but their determination to keep warm and save unnecessary effort is evident in a coiled neck and head that remains nearly lost in an untidy ball of fluffed feathers jogging along above racing legs. If forced to fly they tend to return to their established stretch of river or lake after a short circuit.

Another bird that probably has very distant origins is the Wattled Crane, *Bugeranus carunculatus*, which visits the cold upland lakes and moorlands, where it feeds on the subterranean parts of sedges, and at Giant Root-rat middens. Wattled Cranes also occur on temperate and sub-tropical marshes in southern Africa, and their closest relative is the migratory Siberian Crane, *Grus leucogeranus*. As with the Blue-winged Sheldgoose, this species may have first established itself and had a much more extensive range during the earliest Ice Ages.

That such specialized and distant vagrants should have found niches in Ethiopia is some measure of how very different these uplands are from the rest of Africa. The highlands have undoubtedly acted as an ecological vacuum, drawing in those species that were already well adapted. Even so, most colonists of the highlands have been African species that were sufficiently plastic to cope with the area's many unique distinctions. This adaptability is still evident in some of the common birds of Addis Ababa. Botanically the city is almost wholly exotic, yet the lush flowerbeds of the International Livestock Centre are visited by White-winged Cliff-chats, *Myrmecocichla semirufa*, looking for termites; a pot-hole in a suburban drive resounds to the eccentric rhythm of a Cat Bird, *Parophasma galinieri*, taking its morning bath in the rainwater puddle. On the university campus Black-winged Lovebirds, *Agapornis taranta*, fly in shrieking clusters through the top branches of Australian eucalyptus trees. Footballers on the playing fields ignore the cackling of Wattled Ibis, *Bostrychia carunculata*, circling overhead (Ethiopians have eliminated almost all their large wild mammals but are exceptionally tolerant towards birds: conversation can be drowned by the incessant trumpetings and nasal gurglings of Wattled Ibis nesting beside a house, yet no reprisals are taken). The ibis is one of 4 endemic birds adapted to moist upland marshes or soft soils in exposed, cold areas.

The Thick-billed Raven, *Corvus crassirostris*, typifies the many birds that are simply northern montane isolates of common southern species. Until recently the two ravens would have differed greatly in ecology and diet, with much greater variety for the southern form, *C. albicollis*. By contrast the Ethiopian bird had a more homogeneous environment—a high proportion of its food would have been insects gathered from dung, plant litter and rodent middens

Thick-billed Raven, *Corvus crassirostris*.

Wattled Ibis, *Bostrychia carunculata*

as well as the rodents themselves. The two most pronounced anatomical peculiarities of this bird correspond to its special mode of foraging. Debris is turned over and scattered with the upper arch of the beak while the bird looks between its legs and flips the beak from side to side. Appropriately, the beak is deeper and more curved than that of its less specialized relative, while the black cap and white nape (which are most likely to get soiled by its upside-down digging) are clad in very short, tight-fitting feathers.

The open plateau and rocky areas in which Thick-billed Ravens could earn their living would have been quite restricted in the past. With the arrival of human settlement and forest clearance their choice of food and overall range have expanded. The same can be said for another endemic, the White-collared Pigeon, *Columba albitorques*; both birds are common and widespread.

Plants that have also expanded with the clearing of forests are typified by the aloes. Of the 300 species found in Africa more than 25 occur in Ethiopia and there are extensive erosion surfaces where they are the dominant plant. Aloes tent to hybridize readily, and one effect of removing the forest barriers that formerly separated populations has been to accelerate natural hybridization among aloe species.

Adaptation to local conditions in Ethiopia is most marked in plants. Common throughout tropical Africa, the Scarlet Coral Tree, *Erythrina abyssinica*, has a unique form that only occurs in the south-eastern massif. *Erythrina burana* produces brilliant gamboge blossoms with the first rains and grows on the rapidly drained highland lava flows.

The margins of the south-eastern massif also shelter several very rare and localized birds; all are likely to represent relict populations in last strongholds slowly capitulating to change in their environment and to the advance of one or

Ethiopian aloes. *Top row from left to right: Aloe berhana, A. monticola, A. jacksoni, A. macrocarpa, A. mcloughlinii and A. percrassa. Lower row from left to right: A. camperi, A. adigratana, A. elegans, A. rivai and A. yavellana.*

Eroding pinnacles, Chorchora, Bale

more competitors. The 'age' of these survivors may vary, but because the Ethiopian montane forests are of relatively recent formation, many a non-competitive bird here may quite simply be the victim of continuous immigration or a speedier turnover of species. For example, some of the southern valleys of the Bale Mountains shelter a beautiful white-crested turaco named after an Italian hunter. Prince Ruspoli's Turaco, *Turacus ruspoli*, occupies relatively short middle stretches of these valleys. In those areas where it has the forest to itself the bird uses both broad-leaved and juniper forest, but in those sections that are shared with the White-cheeked Turaco, *T. leucotis*, Ruspoli's is restricted to juniper forest only. Both birds probably belong to the same radiation of smallish, red-winged forest turacos, but the White-cheeked is the most like other common species. It is the regional representative of a type that is found right across Africa. Their very recent spread must have been facilitated and hastened by the ability to inhabit wooded riverbanks in very dry lowlands as well as fully developed moist forests. Ruspoli's has not had the time to become sufficiently distinct and so retreats as its cousin expands.

All Ethiopian relicts are 'new' when compared with those in east Africa, but a progression of ever-drier fringing belts have always surrounded eastern and western Ethiopia. These communities are likely to have been more durable than the true moist forest elements that give every sign of being very recent arrivals. South of the forests where Prince Ruspoli's Turaco still hangs on to existence, the valleys open up, and just before the southernmost margins of the Ethiopian dome get lost in the arid flats below there is a band of open acacia parkland. The area is little known, but it is the focal centre for two unique birds, Stresemann's Bush-crow, *Zavattariornis stresemanni*, and the White-tailed Swallow, *Hirundo megaensis*. These birds are seen further afield from time to time, but why the main population should be found in one small patch of what appears to be fairly nondescript bush country has long puzzled ornithologists. One explanation may lie in the historical interaction between Ethiopia's bird communities and those belonging to neighbouring zones.

While Somali highlands intergrade into the northern foothills of the Bale–Chercher massif without clear definition, the Somali influence declines further to the south. Beyond the Chalbi desert and Lake Turkana (Rudolf) lies east Africa with its rich mixture of communities. Ethiopia's western slopes are an impoverished extension of the vast sub-Saharan savannahs, but few of its representatives cross the Great Rift Valley.

Caught between four major biogeographic zones, each with its own representative communities, the extreme southern tip of Ethiopia exists in a sort of limbo. Some dominant elements from each of the distinct zones just fail to reach

Stresemann's Bush-crow, *Zavattariornis stresemanni*

White-tailed Swallows flying past the Gujuralle peaks, southern Bale massif, Ethiopia.

this area. Take, for example, the crows and starlings. Many species are shared, yet each region contains the core area for much more localized species; ravens and choughs in Ethiopia, White-crowned Starlings in Somalia, various glossy starlings in east Africa. It can hardly be a coincidence that the outermost margins of each regional avifauna either peters out before, or only tentatively overlaps, the Stresemann's Bush-crow's vestigial home. The Bush-crow has much in common with both crows and starlings but is thought to be closest to Eurasian choughs and ground-jays. The ranges of the latter have always lain in a belt south of arctic-influenced uplands and barrens. While Ethiopia was a dislocated African fragment of the great northern ice-cap, its southern slopes would have had much in common with their Eurasian equivalents.

Stresemann's Bush-crow could be in the process of being 'squeezed out' by the slow advance of natural biological influences coming from the south and artificial agricultural influences from the north.

As with other relict species living in stable corners of their former range, survival seems to depend on 'holding off the competition'. New agricultural practices could change the balance and trigger their decline just as they have for choughs in parts of Europe.

The White-tailed Swallow also exists close to the interface between several distinct martin and swallow populations. Here the concept of core areas surrounded by much larger ranges is even more relevant, because most swallows range far and wide but are severely constrained in their nesting needs. If White-tailed Swallows have a peculiar taste in nesting sites (tall termite mounds have been suggested) this might explain an immediate limitation, but ultimate explanations always lie concealed in past vicissitudes of climate and the steady procession of evolving species.

The flora and fauna that Ethiopia shares with tropical Africa tend to be restricted to the most versatile and mobile forms. It was therefore a great surprise in 1978 when two Ethiopian species of the minute earless toads, *Nectophrynoides*, were found, because the only other known localities for this genus were montane forests in Tanzania and Liberia. One species, Malcolm's Earless Toad, *N. malcolmi*, was only found on the Bale plateau between 3200 and 4000 m, where it is very cold and mostly rather dry.

In adaptation to cold, the toads' eggs are fertilized in the female's body, whereas drought is counteracted by the eggs being laid in soil where the mouthless tadpoles feed from a yolk-sac within the egg until they can emerge as miniatures of the adult. Instead of being archaeoendemics, it would seem that these toads (entirely absent from moist lowland Africa) may be specialists that benefited from Ice-Age conditions when lowered temperatures all over Africa combined with drought in all but the most favourable localities. Now eliminated from intervening areas by today's warmer temperatures, these little toads may be part of a very important community whose members might be called 'depository species'. These are species locked into relatively small areas whence they may, one day, be released by the long-term cycling of global climates. When cold and drought return, the toads' unique adaptations should permit them to expand once more—the rare and local may then become common and widespread.

Since the publication of the World Conservation Strategy it has become common to hear animals or plants called 'genetic resources' and any park or reserve an 'on-site conservation unit' for such 'resources'. These clumsy phrases acquire greater meaning with the knowledge that many species await the right combination to unlock their special capacities and release them from some insignificant enclosure. Before they were cleared by people Ethiopia's forests may have been extensive, but they were even then impoverished in species because they were essentially post-glacial in formation. This fragile and vulnerable habitat could

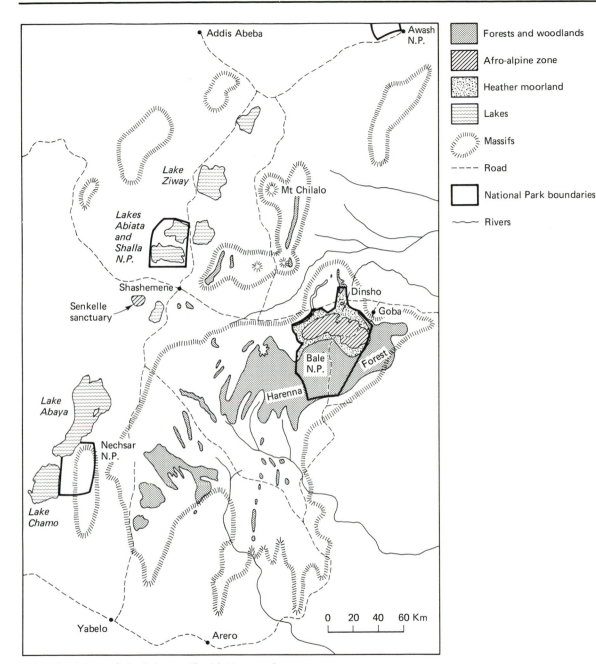

Rift Valley lakes and the Bale massif, with Harenna forest.

scarcely have withstood the severity of Ethiopia's climatic past. Colonization of the forests by species from more robust habitats has been a nearly contemporary process that recapitulates a perennial African dynamic: the movement and exchange of populations between habitats, especially between forest and non-forest.

Ethiopia's innumerable endemics give it a very special significance for conservation, but that realization has been rather late in coming. A UNESCO survey in 1965 led to the establishment of Simen National Park in 1969, and Leslie Brown's visits between 1964 and 1969 alerted the Ethiopian Government as well as the outside world to the need for a vigorous conservation policy. Major architects of conservation in Ethiopia have been Dr Mesfin Wolde-Marian and Com. Teshome Ashine, whose efforts have had the support of several enthusiasts overseas. In 1974, the Pro-Simen Foundation was formed in Switzerland, and its leading figures, Professors B. Nievergelt, B. Messerli and H. Hurni from the Universities of Zürich and Berne, have mobilized resources in Europe in support of Simen.

The single most important conservation area in Ethiopia is the proposed Bale National Park. UNESCO commissioned Dr Bekele to compile a report on Bale; this culminated in the Bale Project in 1981. The Ethiopian Wildlife Conservation Organization then invited the New York Zoological Society to assist, and their representative Dr C. Hillman initiated various research programmes and prepared Bale's first management plan in 1986. Although Bale remains ungazetted, the concept of a national park has received such wholehearted support from senior Government officials, especially the First Secretary for the region, Com. Gezahegne Worke, that other authorities soon came to respect the boundaries and politics of the proposed national park.

Conservation in Ethiopia has seen exceptional progress under a wildlife conservation organization that is exemplary in Africa. Its manager, Com. Teshone Ashine, and Com. Mahammad Abdi have virtually suppressed all trade in wildlife products such as ivory and colobus skins (as an example of the scale of this trade a single dealer took out 26,000 colobus skins in one shipment in 1973). The organization maintains an effective system of national reserves and sanctuaries.

More than most other African countries, Ethiopia has a long tradition of using indigenous plants. The endemic Tef, *Eragrostis tef*, provides the cereal for 'injara', a staple food in the highlands. The Ethiopian Banana *Ensete ventricosa,* or 'Inset' is a plant widely cultivated for its shoots and stem pulp. Safflower or Suf, *Carthamus tinctorius*, is grown for its seeds, which provide a cooking or lamp oil or, when roasted, a flour to thicken and flavour water as a milk-like fasting drink. The Oromo Potato, *Plectranthus edulis*, grows naturally beside streams and is now widely cultivated in the south-west as a major food. 'Sensel', *Adhatoda schimperiana*, has numerous uses; the strong-smelling leaves are still used to treat malaria, as a soapy cleanser and to scour and flavour tela beer pots. This useful plant is commonly used as a hedge.

Many other plants feature in traditional medicine and crafts, but the country that gave the world coffee has much cause to ponder on its natural resources. Berries that had been appreciated by Ethiopians for 2000 years were virtually unknown in London in 1600, yet there were 3000 coffee houses there by 1675. The taste grew, until today coffee is the single largest import of the USA. All this from a scruffy-looking understorey bush hidden away in the lower reaches of Ethiopia's mountain forests.

Ethiopia conceals other treasures as yet unknown to the outside world, but many will only reveal themselves when people have learnt to value an ecosystem more highly than the odd commodity that it produces.

Coffee, *Coffea arabica*, in blossom.

CHAPTER 9

Mountains Old and New— The Equatorial Highlands

Like passengers in a boat, airborne travellers over Africa can look down and forget that another landscape lies beneath the sea of clouds. In common with other mariners, their landmarks will be islands that rise out of the slow frothy surf, each stained, as if by tides, with delicately tinted bands of vegetation. Carried on winds that mimic currents, the clouds swirl and eddy far below, but the aerial observer, like the mountain-tops, is detached. The air is tiered and above 4000 m there are dry cloudless winds that reinforce the difference between mountain-tops and land below. On the tallest peaks these winds can suck up moisture and shed it as snow. The thin air and often cloudless sky above let the upper slopes freeze by night and scorch by day, so it is rare to find tropical organisms that can survive the daily ordeal of sun and ice.

The tallest mountains are nearly all relatively recent volcanoes, but for every soaring cone there are several broken-down stubs, the remnants of older mountains, whose soft pumice ash and brittle lavas have long ago disintegrated beneath the onslaught of solar blow-torching, frosts, ice, rain and wind. With every drift and flexure Africa has thrown up folds and faults, and at every point of pressure volcanoes have erupted. For more than 30 million years the continent's spine has been peppered with cones and craters, but because volcanic materials tend to break down and wash away there are now plains where mighty cones pierced the clouds. Mount Napak in Uganda and Rusinga Island in Lake Victoria are the meagre vestiges of mountains that once reached higher than Kilimanjaro. Mount Mlanje in Malawi and the Aberdares in Kenya are even older and belong to land masses that began to build up in the Jurassic.

Moss Forest

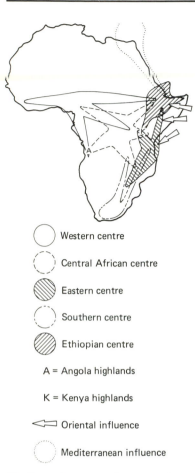

○ Western centre

⌐⌐⌐ Central African centre

▨ Eastern centre

⌐⌐⌐ Southern centre

▨ Ethiopian centre

A = Angola highlands

K = Kenya highlands

◁ Oriental influence

⌐⌐⌐ Mediterranean influence

The five major centres for mountain
communities, with their relatedness to
other centres suggested by polyps.
Mediterranean and Oriental influences
are also indicated.

Each new mountain has spawned its own new life and each geological extinction and levelling has likewise brought its own slow obliteration of organic communities, but the processes, both organic and inorganic, are as contemporary as they are ancient—on the Virunga volcanoes in central Africa it is possible to see the very first colonization of cinders by lichens and ferns, while in Karamoja goats hasten the departure of a last mountain juniper dying on a vestigial volcanic scree.

The flora and fauna of Africa's montane islands are different from the surrounding lowlands. What is more, the 'tide lines' mark out zones that are often quite sharply different from each other. Each mountain differs from every other to a greater or lesser degree. Indeed, like the Galápagos Islands, almost every one can boast some species or subspecies that is all its own, but, like archipelagos of islands, the mountains also form five regional clusters. These mountains, comprise Zanj (or 'the eastern arc'), the central African (or western rift), Cameroon, Ethiopia and southern Africa. There are distant outcroppings in the Sahara, in Nimba and Loma in the far west, the Arias mountains in Namibia and the Bie plateau in Angola, but these have relatively few endemic forms. Volcanoes that have erupted in between these clusters tend to draw their colonists from the nearest source (such as Kilimanjaro from the Usambaras), but there is an anomaly in the Kenya mountains having even greater affiliations with more distant Zaïre than with Zanj, which is closer. Mount Elgon especially seems to have drawn most of its fauna from the west. The maps on this page show the higher montane regions of Africa and their very varied affinities; the

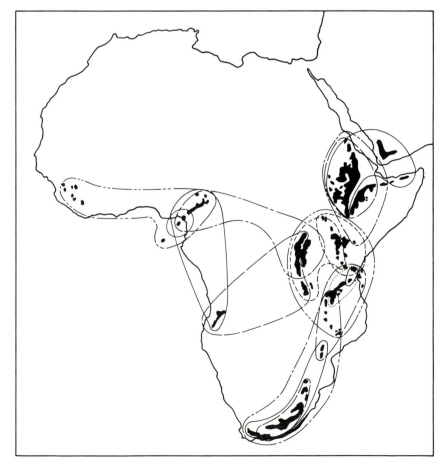

The montane regions and communities
of Africa, with various greater and
lesser affinities indicated.

second map summarizes these relationships in a simplified schematic form. The conformation is a summary of data drawn from many sources. The detailed pattern differs for birds, frogs and balsams, no pattern is quite the same.

Rainfall and temperatures are obvious determinants of fauna and flora, and of their adaptations, but the other dominating influences on mountain life can be reduced to three all-pervasive factors. The first is linked to altitude—how high (and how extensive) are the upper reaches? The second influence is age—how old is the mountain? The third determinant of a mountain's unique biology is its position on the continent.

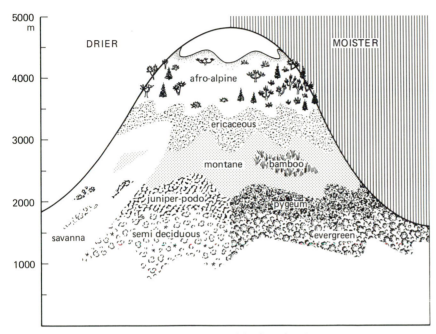

Montane and forest vegetation in relation to altitude and moisture.

The height of a mountain determines the range of habitats it will support. Broadly speaking, there are three altitudinal belts: an Afro-alpine zone, a heath or sub-alpine and a montane forest or grassland belt. Depending on the position and size of the mountains there will be further subdivision of each belt on its drier and wetter faces.

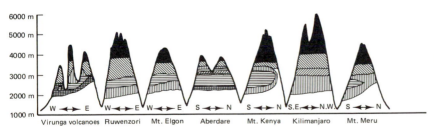

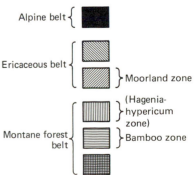

Schematic profiles showing the vegetation zones on the seven highest mountains of East Africa. The wettest side of each mountain is on the left, the letters at their base show the orientation. (After Hedberg, 1951).

The five regions differ greatly in extent, and within each the mountains are of various ages. The eastern arc mountains stretch from Kenya through Tanzania to Malawi and Mozambique and they embrace some of the oldest and newest

4m

Alpine Zone, *Lobelia, Carex, Senecio.*

12m

Subalpine or Ericaceous Zone, *Erica aborea.*

18m

Bamboo, *Arundinaria.*

40m

Montane Forest, *Podocarpus, Cyathea, Ocotea Aningeria.*

mountains on the continent. Another very long chain of old and new massifs marks the course of the western rift. There is a very tenuous link across the Manica and Baluena hills between central Africa and the Angolan highlands. Here the Bie plateau only rises to 2600 m but drops away quite steeply to an arid Atlantic coast; this ecologically stratified escarpment shelters several unique species. The Angola escarpment demonstrates the importance of narrow interfaces between sharply different habitats. Here populations are both isolated and conserved within a series of parallel narrow strips.

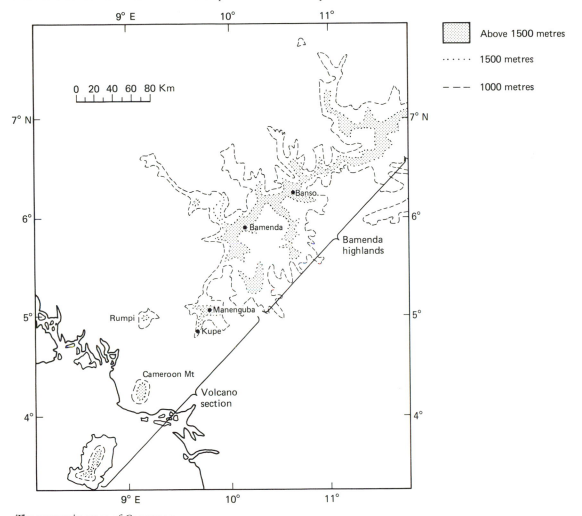

The mountain areas of Cameroon.

The Cameroon highlands include volcanics of various ages. One tall and recently erupted volcano, Mt Cameroon, stands on the mainland and a second, Santa Isobel, dominates the island of Bioko (Fernando Po). A third, Mt Manenguba, is older and marks the southern flank of much more extensive non-volcanic uplands that stretch 500 km to the north-east to peter out in dry savannahs and grasslands.

Outside the tropics, highlands are an integral part of the Cape–Maputo complex that was discussed earlier. Likewise the rather untropical history of Ethiopia has already been examined.

It has already been shown that Africa's southern, eastern and western shores have each fostered their own unique communities. Mountains are, in a sense, inland shores where airborne currents from distant seas shed their moisture to sustain a series of graded shorelines, each another tier of life with its own needs and preferences. Thus the tide-marks on Africa's mountains contain within them layered strata of unique forms. The lower reaches show the most evidence of local 'off-shore' origins. For example, chameleons (a southern and eastern speciality) have diversified into the unique *Chamaeleo schubotzi* on Mt Kenya and *Rhampholeon temporalis* in the Usambaras. Cameroon, a major 'squirrel centre' in Africa, has stranded a very localized species, *Aethosciurus cooperi*, in the Cameroon mountain forests. Three or four species of mountain sunbirds on Ruwenzori and the Virunga volcanoes clearly share common ancestors with the 20 or so sunbird species that populate central Africa.

Fewer off-shore species wash up on to the upper reaches of tall mountains. When recently active volcanoes were first found to have many plants in common with Palaearctic Eurasia and the Himalayas the only explanations seemed to be long-distance colonization or the existence of very recent 'cool corridors'. The fact that a long series of enormous volcanoes have come and gone in eastern Africa for some 30 million years means that alpine flora have actually had an immense time to develop their distinctive African mountain-top forms. As many of these volcanoes have been strung along a chain of Rift Valley uplands, the colonization of new mountains from older ones would have been facilitated, particularly during cool periods.

Two of the most obvious Afro-alpine plants, giant lobelias and giant senecios, derive from lower levels of the mountain and have adapted to ever harsher climates. The giant senecios, which grow very slowly and may flower at 10 to 20 year intervals have none the less speciated rapidly on volcanoes with *Dendrosenecio adnivalis* on the Virunga volcanoes, the closely related *D. kahuzicus* on Mt Kahuzi and *D. erici-rosenii* on Ruwenzori.

Giant senecios belong to the Composites, an old family which has been particularly successful at temperate latitudes and on high cold mountains because the growth pattern of leaves and buds growing directly off the stem allows a very flexible remodelling of the plants' architecture to withstand temperature extremes. For example, leaf rosettes, like feathers on a bird, insulate the inner organism, its vulnerable stem and buds, from frost (giant senecio buds further protect themselves by producing a thick slime that acts as anti-freeze). Giant senecios tend to be most similar at lower altitudes, where they are inconspicuous within a mass of other vegetation. The most distinct subspecies occur on separate mountain-tops, each having derived independently from the populations of senecios below. Subspecies of *Dendrosenecio adnivalis*, and closely related species,

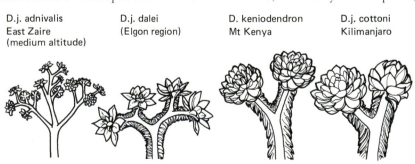

D.j. adnivalis   D.j. dalei   D. keniodendron   D.j. cottoni
East Zaire   (Elgon region)   Mt Kenya   Kilimanjaro
(medium altitude)

The beginnings of mountain (island) evolution. *Dendrosenecio johnstoni* (and related *D. keniodendron*) from four localities (from west to east, but also wetter to drier, warmer to cooler and lower to higher). Note the enlarged rosettes, on fewer thicker stems and increased insulation from dense layers of dead leaves that accumulate on the stem.

illustrate adaptation to higher altitudes, less rain and lower temperatures. Those growing on colder, drier mountains not only have shorter and thicker stems, fewer branches and rosettes; the rot- and pest-resistant leaves are larger as well as being retained on the stem long after they have withered and dried out. Thus the best-insulated and largest race is *D. adnivalis cottoni* at 5000 m on Mt Kilimanjaro. Other plants alter their growth form with altitude, thus *Alchemilla subnivalis* grows erect on branching stems below the frost line but forms dense cushions above. Cushions, rosettes, tussocks and hardened tissues are typical Afro-alpine plant forms. The giant lobelias have also speciated, with the wet-adapted *Lobelia wollastoni* in the central African mountains and the dry-adapted *L. telekii* on the east African mountains. Another, *L. deckenii*, has six different subspecies on six different mountain massifs, a radiation that may have occurred in less than 1 million years. Other plants evolve much more slowly; for example, *Arabis alpina* shows no variation over a vast and discontinuous range.

Close affinities in a small number of species in the Afro-alpine vegetation of the tallest peaks contrast with a great diversity of origins and relationships at lower levels. In spite of Ruwenzori and the Virungas being a lot wetter than Mts Kenya and Kilimanjaro, their upper reaches share many life-forms (although the species tend to differ). Lower down affinities are more parochial. Mount Elgon, an older, corroding volcano, shows particularly strong contrasts between its dry north-eastern and wetter south-western faces. Some of the former are typically eastern species while the latter tend to have western affinities.

Because vulcanism is the major land-building force most of Africa's tallest mountains are volcanic, and because they soon break down they tend, in geological terms, to be very new. Despite their fertile soils, the newness of volcanoes makes them biologically impoverished compared with older and lower crystalline mountains nearby. Yet because they push up so much higher they have climates and foster flora that are intrinsically more different and more specialized than anything lower down. The result is that the peculiar animal and plant communities of Africa's mountains can be categorized into small classes of highly specialized, often recent, colonists, which tend to live on very high well-separated Afro-alpine mountain-tops and much larger and more diversified classes of montane communities, some of which are of very great age.

The greatest number of species come from montane forest but there is also a significant endemic flora in the grasslands, moorlands and other non-forested areas of the mountains, notably in montane Cameroon. The majority of species living in Afro-alpine habitats are likely to be classic neo-endemics because they are recent colonists. Even so, they may be very specialized because selection for survival in such extreme conditions ensures exceptionally rapid adaptation. Thus, there are a number of wingless flies sheltering in the dense leaves and bracts of some Afro-alpine lobelias and senecios. Because they occur on the most recent volcanoes they are likely to be the product of very rapid evolution.

The best-known mountain in Africa is also one of the newest. Compared to the much lower Usambaras, Kilimanjaro is very impoverished—even so, there are a number of minor endemics such as a pretty balsam, the 4 m high Blue Bog Lobelia, *L. deckenii*, and a pale blue gentian, *Swertea kilimandjarica*. The only endemic mammal on Kilimanjaro is a high-altitude subterranean shrew, *Myosorex zinki*.

Kilimanjaro is three mountains in one. Shira, in the west, is the remnant of a volcano that died half a million years ago; Mawenzi is the jagged skeleton of a second giant; while in between Kibo dwarfs both beneath its vast dome of lava and ice.

Mawenzi may or may not have been blasted by Kibo's most extensive eruption 36,000 years ago, but the Chagga people, who have lived in the foothills for over 200 years, have a very geological-sounding legend. One day Kibo was

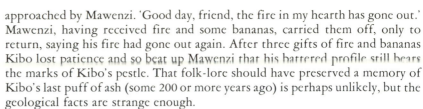

Lobelia and rocks

approached by Mawenzi. 'Good day, friend, the fire in my hearth has gone out.' Mawenzi, having received fire and some bananas, carried them off, only to return, saying his fire had gone out again. After three gifts of fire and bananas Kibo lost patience and so beat up Mawenzi that his battered profile still bears the marks of Kibo's pestle. That folk-lore should have preserved a memory of Kibo's last puff of ash (some 200 or more years ago) is perhaps unlikely, but the geological facts are strange enough.

An age of just over a million years makes this vast landmark younger than mankind itself. At 5960 m it is today the highest of many hundreds of African volcanoes, yet Kibo has already begun to erode away and Mawenzi is crumbling fast.

The mountain's lower reaches may have been colonized from two different directions. The more versatile of the Usambara flora and fauna have invaded from the east—hyraxes, Abbot's Duiker, numerous trees and ferns. The local people even use one local endemic, *Coleus kilimandscharicus*, as a boundary-hedge plant. From the north and west a few species from the main forest block, such as the Pied Colobus, *Colobus guereza*, have managed to cross substantial gaps.

Mt Kenya giant Lobelia.

Mount Kenya, closer to the central African forests, has rather more western colonists, but like Kilimanjaro is not so new that it is without its own endemics. One of these is a green tree viper, *Atheris desaixi*; another is the Blind Mole-shrew, *Myosorex pollulus* (which fills the niche of golden moles, the common subterranean insectivore on other tropical mountains).

In spite of their unique Afro-alpine communities the best-known east African volcanoes, Kilimanjaro, Meru, Kenya and Elgon, all lie outside the main centres of mountain endemism. This is not true for the Virungas in the western rift, nor for Mt Rungwe, a major volcano that marks the junction of the eastern and western rifts. Tall mountain ranges radiate from this point, east, west and south. They link the Rungwe area, however imperfectly, with the three great montane regions of eastern, central and southern Africa.

Lava fields on Mt Kenya.

North of Lake Malawi, moisture is trapped by this complex of mountains to fall as rain, sometimes over 40 cm in a day and up to 400 cm a year. Rain-shadows in the low-lying Rukwa and Usangu valleys mean very steep climatic gradients between mountains and lowland. The complex geology and soils of this relatively small area give the landscape great variety and beauty.

The mountains' ecological diversity is manifest in a flora of some 3000 species. The diverse affinities of these plants reflect the fact that three major mountain archipelagos converge at this point. Small and fragmented montane forests account for some of the interesting plant species, but it is the more

Alpine Mt Kenya

extensive montane grasslands that put on one of the great floral spectacles of the world. Between December and May the Kipengere range, and especially its central Kitulo plateau, become covered in flowers: ground orchids, red-hot pokers, everlasting flowers, geraniums, asters and many more—'Bustani ya Mungu' say the locals, 'God's garden'.

This region may in fact represent an evolutionary centre and pivot for the dispersal of several important flower groups. Among them are the honey-peas, *Kotschya*, and the ground orchids, *Habenaria* and *Halothrix*. These and many more have local species found nowhere else. Links with southern Africa are evident in a local abundance of African iris or *Moraea* species.

Several animals from this particularly wet locality are very dark in colour; there is a deep brown elephant shrew, a melanic lark and a very beautiful squirrel, *Paraxerus lucifer*, with bright red limbs, face and tail.

The honey-pea *Kotschia*

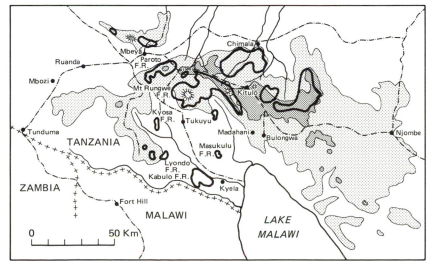

▨	Land above 18000 m
░	Land above 1200 m
⬭	Forest reserves
——	Rivers
—·—·—	Roads
+++	International borders

The southern Highlands centre of endemism at the northern end of Lake Malawi.

A forest of Giant Heather *Erica arborea*, draped in Usnia lichen.

Extensive and complex as they are, the east African mountains are exceeded in biological and geological variety by the highlands of central Africa. In a continent rich in spectacular vistas, magnificent animals and plants, the landscapes of Kivu and Ruwenzori are on a scale and of a beauty that is beyond description. The sky is never still, an ever-changing repertoire of animal noise fills days and nights, and everywhere there is the evidence of extraordinary fecundity.

The highlands of eastern Zaïre stretch 1500 km from north to south. The main part of this long chain is made up of non-volcanic Rift Wall mountains such as the Migula, Itombwe and Ruwenzori mountains, reaching 5100 m in height. In the heart of this complex the Virunga volcanoes rise to 4507 m. Here ancient blocks of upthrust bedrock exist close to half a dozen very tall volcanoes. Some are still intermittently active and the town of Goma only just escaped being buried under lava as recently as 1966.

The affinities of the Afro-alpine communities of Ruwenzori and the Virungas are with other high mountains to the east, but some of the montane forest species suggest earlier connections with other fringing heights around the great Zaïre basin, namely Cameroon, and, to a lesser extent, Angola. Some of the affinities that link separate mountain blocks undoubtedly lie in their sharing connections with a larger intervening region such as the Zaïre or Lake Victoria basin. Climatic changes were more likely to strand former inhabitants of a large climatic region on its peripheral uplands than to ferry fragile species along hypothetical corridors. This is not to deny the existence of such corridors but to suggest that it would normally be the dominant and expansive species that would use them, not the rare and vulnerable ones. Mountain communities have many ecologically specialized endemics but they also conserve representatives of the past flora and fauna of the larger region of which they are a part. I suspect that it will be in this shared inheritance that some of the links and affinities between mountain blocks will be found.

The Cameroon and Angolan highlands share with the Usambaras the distinction of being relatively stable localities where moist and dry climates have probably always met without one ever entirely engulfing the other. Many of the

Tanganyika Mountain Squirrel,
Paraxerus lucifer

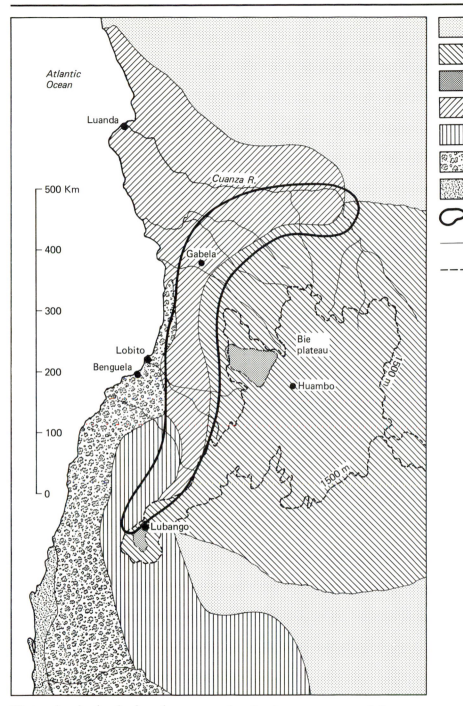

Western Angola, showing how the very steep interface between extremes of climate and vegetation coincides with the centre of endemism.

endemics may be particularly rare and localized because they have become restricted to a rather narrow ecological zone where two climatic extremes meet and interact. The distribution of three endemic birds of the Angolan escarpment conforms with this sort of interpretation (see map above).

These connections are most clearly expressed in the 'pairing' of birds, some of which belong to montane forest, others to more open montane habitats. The

Angola connection mostly concerns forest *edge* species, notably a twinspot finch, a waxbill, two flycatchers and a puff-back shrike. It is likely that each of these rare or localized birds is a relict stranding from some formerly widespread group because, in every instance, they have close relations that are more widely distributed. Even the two or three Angolan endemics that are more obviously distinct (notably a thrush, *Sheppardia gabela*, and the long-bill, *Macrosphenus pulitzeri*) probably have a similar history.

Similar considerations may apply to the montane *forest* birds except that it is the intervening forest that shelters their relations. Perhaps the two groups owe their origins to different dry and wet phases in the past. The mountain francolins have distinct species in each of the three regions (with other species further east); all large, noisy, ground-hugging fowl (see map on p. 29).

The other threesome concerns the rare turacos, *Turaco johnstoni* in central Africa, *T. erythrolophus* in Angola and *T. bannermani* in Cameroon. All three are clearly relict species. They belong to a group of relatively large fruit-eating birds with dextrous legs and agile bodies. Their long tails and broad rounded wings are frequently employed in jerky and flamboyantly ritualized gestures combined with loud choral accompaniments. There are some 20 species occupying the various woodlands and forests of Africa and they show great variety in size and colour, in the shape of beaks and the form of head feathering. The last is particularly ornate in the forest species, most of which have brilliant crimson primary feathers on the wings. Bannerman's Turaco was once called *'Proturaco'* because its coarse undifferentiated beak has none of the special blades, notches or coloured 'shields' of other species, nor are its facial feathers permed into elaborate shapes.

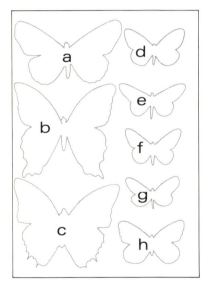

Opposite page. Some endemic butterflies from the Ethiopian highlands. a, Oscars Wanderer, *Acraea oscari.* b, Ethiopian Swallowtail, *Papilio aethiops.* c, Apollo's Charaxes, *Charaxes phoebus.* d, Latticed Wanderer, *Acraea guichardi.* e, Orange-banded Wanderer, *Acraea ungemachi.* f, Clean Wanderer, *Acraea safie.* g, Water Wanderer, *Acraea maji.* h, Ethiopian Dotted Border, *Mylotris mortoni.*

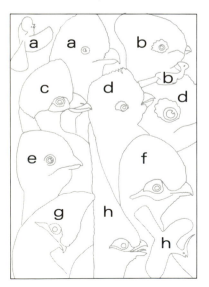

Overleaf. The rare Turacos. a, Bannerman's Turaco, *Tauraco bannermanni,* b, Ruspoli's Turaco, *T. ruspolii.* c, Ruwenzori Turaco, *T. johnstoni,* d, Chercher White-cheeked Turaco, *T. leucotis donaldsoni.* e, Pauline Turaco, *T. erythrolophus,* f, Zanzibar Turaco, *T. fischeri zanzibaricus.* g, Knysna Lourie, *T. corythaix,* h, Mbulu Long Crested Turaco, *T. schalowi chalcolophus.*

The Ruwenzori Turaco does have a more ornate arrangement of facial plumage, but it combines the characteristics of at least three lineages and so could also be the survivor of another basal turaco group. This species has two races. The southern one seems to be a very scarce bird, but the Ruwenzori race, by contrast, is fairly common, albeit within a narrow belt.

Other 'pairs' have speciated in Cameroon and central Africa. Thus, the very colourful Doherty's Bush-shrike, *Malaconotus doherti*, the Olive-backed Finch, *Nesocharis ansorgei*, and more than one peculiar weaver-bird are central African montane birds with relatives in the Cameroon mountains—the Kupe Bush-shrike, *Malaconotus kupeensis*, only known from one isolated mountain forest, must be one of the rarest birds in Africa. Shelley's Olive-backed Finch, *Nesocharis shelleyi*, is a minute finch which unlike its marsh-living, seed-eating eastern relative feeds on insects in the trees. Such versatility in diet may be made easier by the minuscule size of the bird, but it does seem that animals sometimes have greater scope in the mountains for escaping the restrictions of a generic niche. A seed-eater can more readily become omnivorous and a savannah animal can enter forest. This shifting of the niche is probably made easier by the existence of numerous narrow but clearly defined horizontal bands of vegetation growing under very different climates (with further differentials on drier northern faces and wetter southern faces). These narrow altitudinal belts of specialized vegetation types will not contain large specialized animals, but they do offer opportunities for micro-adaptation.

Sunbirds, like white-eyes on oceanic islands, speciate readily on tropical mountains, because there are numerous ways of making a living from flowers and their associated insects. The broad-band species range widely through more than one vegetation type, relying on their own specialized feeding techniques. Others become linked with a particular flora and such sunbirds can be 'stacked' on tropical mountains (although there may be some seasonal overlap and many species tend to move down the mountains during the colder months). The central African mountains have several such birds. The lower margins of the Ruwenzori montane forest are favoured by the Green-headed Sunbird, *Nectarinia*

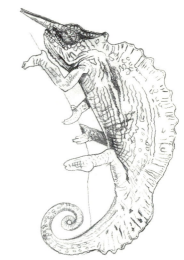

Cameroon Two-horned Chameleon, *Chameleo montium*.

Previous page, top left. Broadbills. *Top left*, African Broadbill, *Smithornis capensis*, *top centre*, Grey-headed Broadbill, *Smithornis sharpei*, *top right*, Himalayan Broadbill, *Serilophus lunatus*. *Middle left*, Green Broadbill, *Pseudocalyptomena graueri*, *middle right*, Rufous-sided Broadbill, *Smithornis rufolateralis*. *Bottom*, Bornean Emerald Broadbill, *Calyptomena hosei*.

Previous page, top right. Golden Monkey, *Cercopithecus mitis kandti*, from the Virunga volcanoes.

Previous page, bottom. Orchids from the southern highlands of Tanzania. *Top left*, Habenaria occlusa, *top right*, Disa longilabris. *Centre*, Disa ukingensis. *Lower left*, Satyrium comptum, *lower right*, Holothrix hydra.

Opposite page, top. Cameroon two horned chameleon, *Chamaeleo montana*.

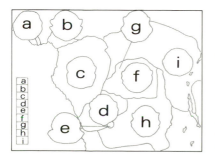

Opposite page, bottom. Distribution of the 'cephus' or 'lesser peripheral' group of monkeys in the central forest block. *a*, Sclater's Monkey *Cercopithecus sclateri* and *b*, the related guenon; *C. erythrotis*. *g*, a newly observed and as yet unnamed type, which may be a variant of Whiteside's Redtail *C. ascaniua whitesidei*, *f*, but suggests an earlier connection between *erythrotis* and *ascanius*, subsequently interrupted by the expansion of moustached guenons, *C. cephus*, *c*. *i*, Redtail *C. ascanius schmidti*. *e*, Lunda Redtail *C. ascanius atrinasus*, *d*, Western Redtail *C. ascanius ascanius*, *h*, Katanga Redtail *C. ascanius katangae*.

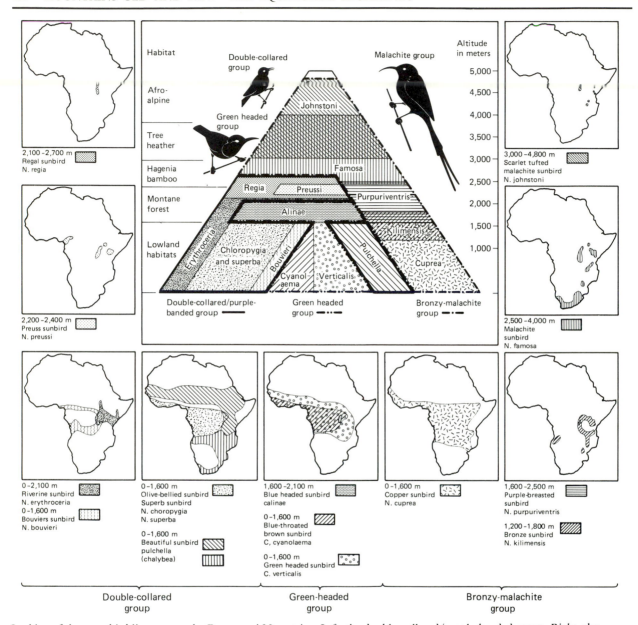

Stacking of three sunbird lineages on the Ruwenzori Mountains. *Left*, the double-collared/purple-banded group. *Right*, the bronzy-malachite group. *Centre*, the green-headed group. Distributions are consistent with the older populations within each lineage having become progressively more and more restricted by subsequent (more advanced) lowland populations.

alinae. The Purple-breasted Sunbird, *N. purpureiventris*, is most commonly seen between 1500 and 2500 m (it is probably rather mobile within this band). Higher up, where landslides and fires generate rich herbaceous growth (between 2100 and 2700 m), the gem-like Regal Sunbird, *N. regius*, is common. In alpine east and central Africa the Scarlet-tufted Malachite Sunbird, *N. johnstoni*, seldom descends below 3000 m. Preuss' Double-collared Sunbird, *N. preussi*, is common to both Cameroon and the central and east African highlands, but in the east slots into a narrower band (below 2400 m). On Mt Cameroon Ursula's Sunbird, *N. ursulae*, is essentially a montane forest colonist from a southern savannah stock of sunbirds.

The examples of non-forest finches and sunbirds colonizing the montane for-

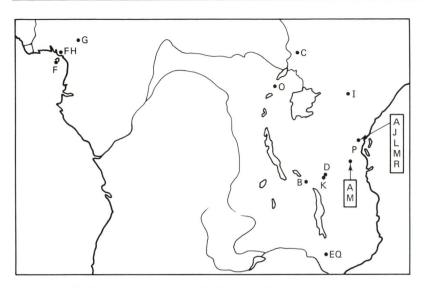

A   Chameleo deremensis
B   Chameleo fuellebarni
C   Chameleo kinetensis
D   Chameleo laterispinis
E   Chameleo mlanjensis
F   Chameleo montium
G   Chameleo pfefferi
H   Chameleo quadricornis
I   Chameleo schubolzi
J   Chameleo spinosus
K   Chameleo tempeli
L   Chameleo tenuis
M   Chameleo werneri
N   Chameleo wiedersheimi
O   Chameleo xenorhinus
P   Rhampholeon brevicaudatus
Q   Rhampholeon platyceps
R   Rhampholeon temporalis

Highly localized chameleons in tropical Africa (mainly montane).

ests can be augmented with others drawn from quite different taxa. That essentially south-eastern and African savannah reptile, the chameleon, has colonized Ruwenzori (*Chameleo xenorhinus*) and also Cameroon (*C. quadricornis, C. montium* and *C. pfefferi*) and there is much to suggest that forests have, for significant periods of time, been so diminished that much of Africa's forest fauna, both lowland and montane, had to be recruited from non-forest sources.

Among birds, some of the most intriguing colonists seem to have come in from outside Africa. South-east Asia with its much more extensive forests may be a very long way away, but a few robust and adaptable vagrants may well have helped repopulate Africa's forests during periods of forest expansion. Since these expansions and contractions are known to have taken place many times, Africa's forest flora and fauna can be described as being deeply laminated, with each of the many layers of species representing different periods of adaptation. The special interest of African mountains is the way they may trap samples of such earlier populations within their own ecological layering.

One of the most interesting of these is the African Bay Owl *Phodilus prigoginei*. Before 1952 when this bird was first described, bay owls were only known from a limited area of south-east Asia. They are a very primitive form of barn owl, which may owe its worldwide distribution to a very refined technique of acoustic prey-detection. Because they roost in caves this group of owls has a good fossil record which goes back 25 million years. Before the improved type of barn owl emerged, bay owls may well have ranged nearly as widely and with as little variation as their successors; certainly the surviving African and Oriental species are astonishingly similar considering how far apart they are. The special interest of these relicts is that the fossils hint at a great age and stability for the refuges where these birds are found. Lake Lungwe in the Itombwe Mountains of eastern Zaïre is where the one and only specimen was collected and this locality is known to be very old geologically and could therefore have enjoyed an environmental stability that goes back tens of millions of years. As if to reinforce this claim, Africa's most primitive shrew, *Myosorex schalleri*, is also known from this region, also from a single specimen, and its pedigree may be even older than that of the bay owls. The near certainty that bay owls were originally without a special environment also suggests that we should try to analyse what it is about these mountain habitats that is so beneficial to conserving species. Great age and stability are obvious components but much more will need to be learnt

Congo Bay Owl, *Phodilus prigoginei* (above) and Albertine Owlet, *Glaucidium albertinum* (below) from the Kivu region of Zaire.

before we can define precisely those combinations of factors that enable the competition to be held off by so many species.

The Itombwe Mountains have further Oriental connections, notably the cuckoo-shrikes, which are predominantly Australian, Asiatic and Madagascan birds and probably latecomers to Africa. Grauer's Cuckoo-shrike, *Coracina graueri*, of the central African mountains is more like its Oriental relatives than any other African species. It is reasonable to suppose that it may be both the relict of an early immigration as well as sharing ancestry with the Azure Cuckoo-shrike, *Coracina azurea*, which is widespread in lowland forest. Here is a separation into montane and lowland forms with the strong implication that the montane form is older.

Another tantalizing connection with Asia is demonstrated by a small group of primitive passerines, the broadbills (Eurylaimidae). These are small sedentary birds which behave like sluggish flycatchers, sitting still in the undergrowth for long periods, flying out to catch a passing insect on the wing or off the forest floor. They probably rely on a steady year-long supply of insects within a small home area. There are ten species in south-east Asia and three in forested Africa. All three occur in eastern Zaïre and western Uganda where a fourth species, one of the rarest of African birds, the Green Broadbill, *Pseudocalyptomena graueri*, only lives in forest close to the Rift wall. Lord Rothschild, who first described this bird, thought it was a flycatcher and therefore thought its resemblance to *Calyptomena*, a south-east Asian genus, was due to convergence. Since then the resemblances have been found to have a real and close genetic basis. The time of the broadbills' entry into Africa is impossible to guess, but the Green Broadbill is perhaps the oldest form, now in the last stage of having been 'squeezed' by its relatives.

There is some overlap with the Grey-headed Broadbill, *Smithornis sharpei*, but this larger bird favours low undergrowth at lower altitudes, whereas Green Broadbills live a slightly more exposed life, snatching insects along the margins of clearings and breaks in the montane forest canopy. Although the grey-headed birds live in low-altitude forest, their distribution is restricted and they too seem to be relatively feeble colonists.

The Red-sided Broadbill, *S. rufolateralis*, smaller than the Green Broadbill, is a lowland forest bird. From Uganda across to Sierra Leone one of the most familiar noises of the forest is the dawn and dusk croaking of this little bird. It is made by males during a one-second circular flight from a fixed perch, and this spasmodic but effective self-advertisement could be one of the ways in which a sombre, sluggish forest-dweller can improve its chances of mating and achieve a more even spacing of males. (That the calls are similar to amplified frog croaks could be a convergence in good acoustic design or perhaps a ploy against predators.)

These considerations apply equally to a fourth species, the African Broadbill, *S. capensis*, which has an even louder croak and is common in riverine, montane and other forests from Natal to Guinea. It is of the same size and has very similar habits to the very much quieter Green Broadbill. Is the African responsible for the Green's decline? Is the Green's inferiority primarily ecological or behavioural? Could amplified communication be decisive for the two competitors? Will a shrinking range bring with it a *coup de grâce*, not from man, but from a noisier cousin?

The broadbills may look insignificant, but potentially they could illuminate several vital evolutionary processes, and the only places such studies could be made are in shrinking forests close to the Great Rift. The Itombwe mountains support the largest remaining populations of Green Broadbills. Overlooking the northern tip of Lake Tanganyika these mountains have the most diversified terrain in the region and a very rich fauna and flora: 565 species of birds have been recorded there by Dr A. Prigogine, an ornithologist who among many interesting discoveries made the first and only record of the African Bay Owl. Maps of proposed conservation areas in the Itombwe mountains are shown on pp. 188 and 274. The Itombwe mountains shelter other unique endemics such as the forest ground thrush, *Turdus oberlaenderi*, Chapin's Flycatcher, *Muscicapa lendu*, and Rockefeller's Sunbird, *Nectarinia rockefelleri*. Frogs include the specialized Bamboo Frog *Callixalus pictus* that breeds in the water-filled hollows of broken bamboo stalks and a giant torrent frog *Phrynobatrachus asper* (one of three endemic torrent frogs) and the Copper-coloured Treefrog *Chrysobatrachus cupreonites*. This species may be a locally evolved special offshoot of an advanced type of treefrog, *Hyperolius*, a reminder that such mountains are laboratories for new species as well as refuges for old ones. The Chisel-toothed Shrew *Paracrocidura graueri* may be just such a local specialist as may the local chameleon *Chameleo graueri*.

Rare endemic species occur in all the major mountain blocks fringing the Zaïre basin and they are drawn from a very wide variety of plant and animal types.

Cameroon has numerous forest primates, bats, rodents and duikers, but when it comes to rare species that are demonstrably primitive Ruwenzori and Kivu have the larger share. For example, there is a dwarf otter-shrew, *Micropotamogale ruwenzorii*, an aquatic insectivore which lives in the clear waters of pebbly mountain streams. Its nearest relatives are the Tenrecs of Madagascar, a lineage that goes back at least to the Eocene, 50 million years ago, when these and other insectivores were widespread in the Old World. They have been pushed out by other small-scale hunters like mongooses, insectivorous rodents and birds. Yet in or around Ruwenzori, mountain streams flow as they have for millions of years, stable, predictable. Here the dwarf otter-shrew uses its bewhiskered hydrofoil nose in a unique hunting technique, snuffling out small crabs, worms and aquatic insects. More adaptable species may eat comparable prey elsewhere, but on these short reaches of rushing clear water the otter-shrew is unassailable.

Two swallow-tailed butterflies from Central Africa. *Above, Graphium gudenusi* (U) *and below, Papilio leucotina* (L).

As has already been mentioned, the forest streams around Mt Nimba in the far west are home to another species of dwarf otter-shrew, *M. lamottei*. Both these animals were only discovered in recent years, so it is just possible that dwarf otter-shrews may yet turn up in Cameroon. However, the two known species are separated not only by a great distance but also by significant anatomical features so they are likely to have been distinct for several million years.

The Ruwenzori Otter-shrew, *Micropotamogale ruwenzorii*.

Ruwenzori Horse-shoe Bat,
Rhinolophus ruwenzori.

Similarly ancient connections are likely for a group of horseshoe bats which is represented by a scatter of very rare and highly localized species. *Rhinolophus maclaudi* is known from caves in Guinea; its closest relatives are two equally scarce species from Ruwenzori and south of Lake Kivu, *R. ruwenzorii* and *R. hilli.* All are distinguished by the relatively simple structure of their large noseleaf and share this and certain primitive features of skulls and wings with a group of rare horseshoe bats from the Far East. Although nothing is known of just how sedentary or mobile these bats are, they are more likely to resemble the otter-shrew in being specialized relicts rather than long-distance mountain-hoppers. *Rhinolophus* is a very ancient genus, known from fossils that are some 40 million years old. The skull of *R. ruwenzorii* is comparable with these fossil remains.

Understanding the timing and effects of climatic and biological changes in Africa is subject to continuous enquiry and dispute. The discussions are not only academic, restricted to palaeo- or biogeographers; there are wider issues that require an understanding of the evolution, biology and conservation of innumerable African organisms and communities.

Take Gorillas, for example; they are found today in two widely separated regions, Lowland Gorillas of the race *Gorilla gorilla gorilla* in Cameroon and Gabon, and a scatter of populations in the vicinity of Lake Kivu. In recent years saving Mountain Gorillas from extinction has become a worldwide campaign. How relevant or realistic is it to respect racial differences in an ape and attempt to conserve the 300–400 Mountain Gorillas that remain? For many, the very existence and splendour of Mountain Gorillas is eloquent enough, but there are also man-oriented reasons for our concern. A recent study of the chromosomes of man and apes suggests that the Gorilla is genetically closer to man than any other anthropoid; it is a closeness of detail that was previously unsuspected.

In this perspective the Gorilla can be seen as mankind's surrogate in the forests. For a long period of leisurely evolution humans left this rich but difficult habitat to their large herbivorous cousins. We evolved outside the forest and were originally excluded from it by the very adaptations that had transformed us into humans.

The existence of numerous montane specialists proves that high mountains are a challenge needing numerous subtle adaptations. While humans have bypassed this biological adaptation, Gorillas have not. Four hundred Mountain Gorillas are all that remain to prove that a large terrestrial anthropoid did once make the *biological* adaptations necessary for survival in the wet, cold vegetation of these African mountains.

Yet the Gorilla, like some other rare creatures, seems to have had mixed success: present in Kivu–Virunga but absent from Ruwenzori; once abundant in the lowlands of Cameroon but unknown in the mountains. Were they expelled by people at an early date or are there more subtle reasons for their absence? Mountains have been important Gorilla retreats, not just because they are always moist, but also because a combination of landslides, volcanic activity, cyclical phenomena (such as the periodic die-offs of bamboos and other plants), elephants, fires, tree-felling and the Gorillas' own bulldozing have combined to generate the vast quantities of low-level herbage and vines that are necessary to support a Gorilla population. The closing of the forest canopy, for whatever reason, would always carry with it the risk of shading out their food supply and this too could explain the absence of Gorillas from areas that seem well suited to them today. If only lush pastures of quickly renewed green growth will sustain Gorillas the greater part of any tropical forest will be closed to them.

There is also the question of competition. Colonization of the mountains by other herbivores could have been inhibited in the Kivu mountains by the prior arrival and local adaptation of ancestral Mountain Gorillas. As if to prove they can protect their interests, Mountain Gorillas will displace duikers and buffaloes and even stampede elephants off their pastures. The question remains: how long have Gorillas lived here? In his pioneer study of the Mountain Gorilla, the American zoologist George Schaller concluded that all Gorillas had a recent common origin and that those living in the Zaïre lowlands had parted from their mountain relatives at a particularly recent date.

Schaller's suggestion of a recent divergence therefore implies rapid selection, but rates of genetic change in Gorillas are at present unknown and it is equally possible that these differences could be of very long standing. Even in the more homogeneous population of western Gorillas, Colin Groves, a contemporary taxonomist, believes he can recognize four 'demes' and it is interesting that some of these seem to correspond with refuge areas for other forest animals (see fig. on p. 182).

It is quite likely that new techniques may eventually be able to provide an objective measure of just how distinct the various populations of Gorillas are. It is also possible, however, that true Mountain Gorillas may be extirpated less than 100 years after their discovery and long before their relevance to human evolution and a continent's history has been adequately appreciated.

The few hundred square kilometres of African montane forest that Gorillas once occupied is now joining more than 100 million square kilometres of land that is being rapidly colonized by 4750 million humans. There is now so little tropical mountain forest left that its biological distinctness is in danger of disappearing. The mountains are Africa's Galápagos Islands—islands enriched by Golden Monkeys, Gorillas and iridescent sunbirds, by giant lobelias, everlasting flowers, Ruwenzori Turacos and all the questions they raise. They deserve greater recognition, protection and study than they have received so far. Scattered through more than a dozen countries, they also require a global approach where the lessons of one country can be learnt by another.

Zaïre has very extensive areas of montane forest and sub-alpine habitats within well-established national parks. Poaching of elephants and Gorillas was commonplace in the 1970s, but in recent years the Zaïre Institute of Nature

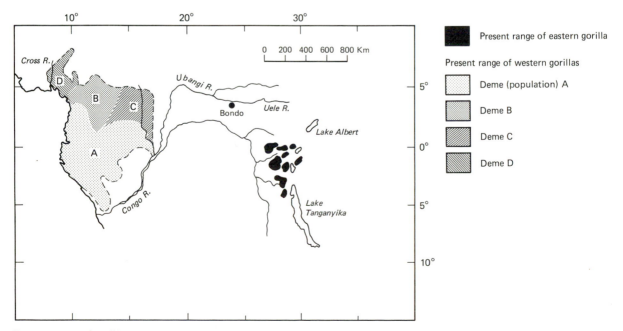

Present range of gorillas.

Delanys Mouse, *Delanymus haymani*

The National Parks in eastern Zaïre, Uganda, Tanzania and Kenya which protect mountain and forest communities.

Opposite, Individuality, age and emotion in the faces of mountain gorillas.

Conservation, with support from the Frankfurt Zoological Society, has commissioned Conrad and Rosalind Aveling to improve the conservation prospects of Gorillas. They have initiated surveys, habituation of some accessible groups and an education campaign in schools and villages. The future conservation of the vitally important Itombwe mountains is under active discussion and the FFPS is supporting a conservation survey of the region.

In Uganda, the Impenetrable Forest is a key locality for the conservation of Mountain Gorillas and a truly representative sample of montane forest flora and fauna because it is of non-volcanic origin. The reserve includes one of a very few high-altitude swamps where the unique Grauer's Swamp Warbler, *Bradypterus graueri*, and Delany's Climbing Mouse, *Delanymus*, occur. Between 1983 and 1986 some 15 Gorillas were killed here and fierce local barons intimidated all efforts at control. Since 1986 a team of disciplined 'Gorilla guards', with support from the new and effective NRA Government, has controlled poaching and a new scientific research station has been built at Ruhiza. The Impenetrable Forest Conservation Project is one of three important regional initiatives to save the Mountain Gorilla that were set up under the auspices of the international Mountain Gorilla Project.

Destruction of montane forest in the central African mountains was greatly accelerated by the World Bank, which financed the felling of large areas in the 1960s and 1970s for pyrethrum farms. These subsequently failed and the Bank's interest in the area waned. Since then earnings from pyrethrum have been far outstripped by revenue from visitors coming to see the remaining Gorillas in Rwanda. It has been suggested that these earnings could be increased if the extensive areas of fallow left over from the pyrethrum clearings were turned back into natural habitats so that some Gorilla populations could regain viable home ranges.

The long-term future of Mountain Gorillas will be bleak unless some effort is made to return some of their recently lost range to them. It has been estimated that Gorillas previously occupied eight to ten times their present range in Kigezi (Uganda) alone.

Their plight illustrates a fundamental weakness in the role of nature conservation in the contemporary world. At all levels—local, national and international—planners will only concede marginal lands to conservation and the boundaries of exploitable land are constantly being rolled back. Most of the remaining Mountain Gorillas have been confined to land that is marginal to agriculture and is often close to marginal for the Gorillas too.

Once able to descend to lower warmer levels during cold spells, todays populations are especially vulnerable to disease because they are denied access to their former refuge areas. In an unusually cold year, many if not all the gorillas could die.

The central African countries are not yet major tourist centres and Gorilla-viewing is still an esoteric taste, yet Rwanda already makes significant earnings from Gorillas, although the animals and their habitats remain 'marginalized' both physically and mentally. Gorillas and the wilderness have only been permitted to exist along the edges of man's physical estate and even more uneasily in the outer reaches of his consciousness.

Yet this is now changing. Whereas an opinion poll in 1975 showed that more than half the population of Rwanda was hostile to Gorillas, another poll 10 years later showed that a large majority had acquired positive feelings towards them and a concerted education campaign had generated pride in Rwanda's custodianship of the Gorillas. There are effective scientific research stations at Uinka in the Rugege forest and studies on the social behaviour and ecology of Mountain Gorillas have intensified at the Karisoke Research Centre, where

Rwandan biologists from the National University have opportunities to carry out higher-degree research.

Even more revolutionary has been the change in perceptions overseas. Grotesque King Kong images are still prevalent, but a growing taste for natural history films has brought Gorilla infants in to frolic on television screens in countless living-rooms. A long-nosed visitor grins nervously into the camera as his appendage is gently tweaked by the gigantic fist of a wild silver-back. A wet and miserable primatologist is cuddled by one of her subjects. Glancing out from below night-black brows, almost lost in the shadows of deep sockets, tentative brown eyes send mute messages of humour, curiosity, affection and *joie-de-vivre*.

Saving Africa's mountain communities is one of the most pressing priorities for world conservation. It is a real test of values and resolve, and Mountain Gorillas are the central challenge.

Join a family, if you can, as they forage through the tangles, pushing, plucking, burping and sighing. Measure yourself beside the physical presence of a silver-back. Is sharing the planet with a few Mountain Gorillas a luxury our squalid civilization cannot afford? Is it a privilege we deserve?

CHAPTER 10

An Evolutionary Whirlpool—
The Zaïre Basin

The ferry which crosses the Zaïre River below its last tributary links the towns of N'Gabe and Kwamouth. It is a stretch of fast-moving river where two columns of water run side by side. The right column has drained dry plateaux not far from the margins of the Sahara, misty volcanoes in central Africa: Lake Tanganyika and the Muchinga mountains in Zambia: some of its sources are 3500 km upstream. Waters in the second column, the Kwa, have travelled 2000 km from the uplands of central Angola.

Downstream from the Kwamouth ferry the sweetened waters of a thousand rivers are mixed in Kimpoko pool and then churned to foam in a giant whirlpool near Luozi.

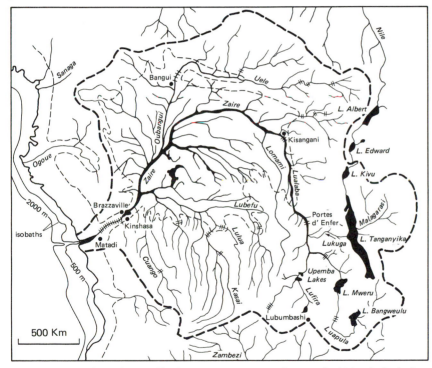

- - - - 500 m
⊣⊢ Falls or rapids
- ▬ - Boundary of basin

Zaire basin with boundaries of basin, 500 m contour and 500 and 2000 m isobaths in Atlantic Ocean.

Only exceeded by the Amazon in size, the Zaïre River dominates Africa—its basin is the huge green belly of a continent. Yet it is a basin that has not always been green. Rain on its distant highland margins has always fed these rivers but there have been many periods when, much diminished, they flowed across arid Kalahari sands that stretched up to meet the dry lands of the southern Sahara. Even then there would have been fringing gallery forests, narrow where the banks were high but expanding into great swamp forests in the sump-lands.

During the most arid periods these riverside forests trailed over a much

Levee forest with numerous lianes.

Congo Weaver, *Euplectes anomala*

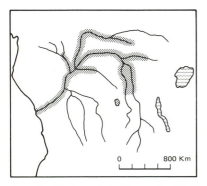

The distribution of the Congo Sunbird *Nectarinia congensis* and the Congo Weaver *Euplectes anomala.*

larger area than the vestigial jungles of Biafra and the central African uplands. At such times riverine galleries became by far the most extensive of all forest habitats in Africa. This is not to say that they would have been homogeneous or continuous — far from it. Even today sluggish trickles over the flat dry plateaux north of the equator are very different from Kivu torrents or sheltered streams off the Angolan uplands. Shallow headwaters are like a multitude of twigs, the deep broad rivers they feed are single great limbs or trunks. Its 'tap root' cuts a channel in the seabed that extends 250 km out to sea and to a depth of 2500 m. With good reason, N'Zaïre means 'the river that swallows all rivers'.

For the plant and animal communities along the banks, differences in river type are matched in significance by contrasts in season. The Zaïre basin spans 22° of latitude. North of the equator rain falls between April and August followed by hot Harmattan winds, whereas the south is cooled by Atlantic breezes and rain from December to March. On the equator two wet and two dry seasons vary greatly in length from year to year and place to place. Both habitat and breeding seasons are affected by these differences.

The great majority of fauna and flora are found throughout the basin, thereby enhancing the impression of a single relatively uniform forest, but subtle hints of past vicissitudes remain beneath today's broad green canopy. Like the reflective surface of Kimpoko pool, uniformity glosses over a multitude of hidden sources. There are actually enough clues (especially in the distribution of some key animals) to begin a reconstruction of the basin's biological history.

Repeated periods of aridity undoubtedly explain the absence of many organisms from the centre of this basin in spite of their occurrence in forests to the west and east, but there are other species such as the Pigmy Chimpanzee and the Crested Mangabey that occur *only* in the central basin. A single river system has promoted the uniform spread of many dominant and versatile species but the same network and regional differences in climate, soils and habitat have helped to fragment others. Organisms of both forest and non-forest origins have been enmeshed and their further evolution has been influenced thereby.

As a theatre of evolution the basin's most peculiar characteristic has been the fact that it is dissected into a network, and these river courses have tended to endure regardless of the dominant climate over the basin itself. The species that have accommodated to this differ from those that are unequivocally forest- or arid-adapted. These would only have crossed or recrossed the basin when the climate had modified the vegetation sufficiently to allow them to advance on a fairly broad front.

By contrast there are numerous species that have undoubted savannah origins that remain enmeshed in the Zaïre net today (a relatively humid period). Likewise many forest species have probably survived repeated epochs of aridity as essentially riverine forest dwellers. Populations of both types would have pulsed in concert with changes of climate, but these pulses would have emanated from drier cells or wetter galleries *within* the basin and on numerous rather than a very few fronts. Some of the possible consequences for evolution will be explored shortly in an examination of the guenon monkeys.

The basin can be subdivided into at least five subregions, each typified by differences in local climate, soil and habitats and to a greater or lesser degree each subregion has become a centre for speciation. Continuous interaction between forest and non-forest habitats has been a major agency in evolution all over Africa, but the form this interaction has taken differs from one region to another. Simple forest/non-forest definitions are often blurred in Zaïre communities, yet study of these very ambiguities should help us understand the process at work.

The savannah types engulfed by the forest include animals and plants from all levels and types of forest. Burrowing toads underground, elephant shrews and rats on the floor, galagos in the undergrowth and birds in the canopy or on the sunny riverbanks. For example, of species in the centre of this forest whose origins are unmistakably outside, one is the weaver-bird, *Euplectes anomala*, while another is the Purple-banded Sunbird, *Nectarinia congensis*. However, both these birds are restricted to the banks of the main river, where they live a relatively exposed life. The weaver fusses about amongst the scarlet dates of the climbing, liana-like rattan palms, shredding leaves for nesting material. Close by, another palm, the Raphia, fountains out its 20-m-long fronds to force open its own place in the sun. Between palms and river the sunbirds flutter around the heavily scented white blossoms of Camwood Trees, *Baphia*, and perch on the columns of giant Ground Orchids, *Eulophia porphyroglossa*, each taller than a man (and another open swampland plant). They remain marginal species, not true forest denizens. Not so the tiger-bittern *Tigriornis leucolophus*, which inhabits swampy rivers under a closed canopy. Although it is also found in west Africa, the Zaïre basin is its main habitat. The climatic swings that repeatedly 'shaded over' Africa's equatorial streams could have made this an original site for adaptation to forest, but this remains uncertain because there are tiger-bitterns elsewhere in the tropics. The type most closely resembling the African species is in Papua New Guinea.

A mammal that ranges widely in dry parts of east Africa but also in Zaïre has distinctly riverine associations. The preferences of the Four-toed Elephant Shrew, *Petrodomus tetradactylus*, are evident throughout its east African range, where it can exist in very arid regions by keeping to the thickets and scrub that

Congo Sunbird, *Nectarinia congensis*.

Okapi, *Okapia johnstoni*

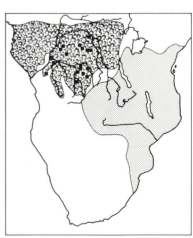

Distribution of Four-toed Elephant Shrew *Petrodromus tetradactylus* in the dry woodlands of south-eastern Africa and in lowland rainforest in central Africa, one of the many indications of past drier climates in the Zaire basin.

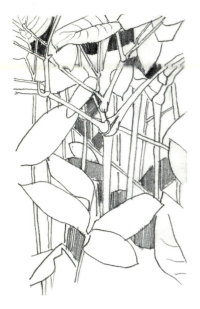

Typical riverine undergrowth,
dominated by arrowroot *Marantochloa*.

fringe seasonal rivers and other drainage lines. In the Zaïre forests this elephant shrew ranges very widely south of the main river while another elephant shrew lives north of the river. Similar adaptations have been made by the Target Rat, *Aethomys longicaudatus*, the only forest dweller of several successful and wide spread bush-rats.

Of larger savannah mammals that have been engulfed by the forest, the best known is the Okapi, *Okapia johnstoni*. It has been established from Pleistocene fossils that Okapis and Okapi-like giraffes once ranged widely through the African moist savannahs, but the fossil record also shows that since then fire has been used more and more by man, while antelopes—the Okapis' potential competitors—have steadily increased in numbers and variety of species.

The Okapi's most basic need is for year-long fresh green growth. Its most basic advantage is the ability to reach branches well over 2 m above ground. Like their cousins, Giraffes, Okapis are very persistent in seeking out nutritious foliage. They walk from bush to bush plucking out only the best green growth with their long muscular tongues. Their browsing is very light, with practically no changes in diet or range over the seasons. In the areas in which they have been studied the dominant trees are of the cassia family, Caesalpiniaceae, and the animals frequent river valleys and clearings or 'chablis' as well as higher ground where they also browse from shade-loving shrubs growing in the forest proper. Ruminant diseases, predators and hunting may also determine in which areas Okapis can or cannot survive. Until relatively recently, Okapis probably found a network of suitable country all the way from south Sudan to Angola, but their

Galago inustus.

principal refuge today is among the minor tributaries of the north-eastern Zaïre basin. Densities can be surprisingly high, with up to one animal per 2.5 sq km and an average home range of 5 sq km for females (male ranges are much more extensive).

Occasionally, diets provide some measure of an animal's origins. Take the species of bush-babies that occur in the main forest block, all of which eat fruit, insects and resin. Although the last has some nutritional value it is mainly a seasonal food, a stop-gap filler that buffers a shortage of fruit and insects in the dry season. The bush-babies vary greatly in their preferences. Predictably, the smallest mostly eat insects, the larger species more fruit, but two bush-babies of the main forest block, *Galago elegantulus* and *G. inustus*, take a very high proportion of gum in their diet. This specialization probably began in the more arid Zaïre basin of the past, but when other bush-babies returned with the rain, yesterday's necessity became today's virtue (or at least a safe ecological niche).

In the eastern Zaïre basin there is a toad, *Hemisus olivaceus*, whose burrowing habits are shared by relatives living further to the south and east. Burrowing is clearly an adaptation for surviving long dry seasons. In the larger rivers, there are other amphibians typical of the savannahs, but exposure and bright light on broad riverbeds probably disqualify these as true forest habitats.

Forests tend to be a dead end for grass-eating animals. What grasses exist are either unpalatable or scarce. Nevertheless the vicissitudes of the Zaïre basin have enclosed several such animals. Two types of acridine grasshoppers, *Odontomelus* and *Holopercna*, have been transformed by forest living. Instead of flying to escape predators, they process plant chemicals, develop brilliant warning colours and exude foul-smelling froths to defend themselves. Scent rather than sound also becomes the means to attract mates, so they have developed long sensitive antennae, stridulate less and use their short wings less. The poor light requires large eyes, and the fact that illumination is both faint and far above may explain why some of these grasshoppers have two pigment zones in their eyes (like graduated dark glasses or sunshades).

Among the many canopy birds that originate outside the forest is the Collared Lovebird, *Agapornis swinderianus*. Lovebirds also illustrate how a bird type that originates outside Africa (they are lories, an essentially Oriental type of parrot) can recolonize a forest habitat after invading Africa across a savannah bridge.

Hardnosed Toad, *Hemisus olivaceus*.

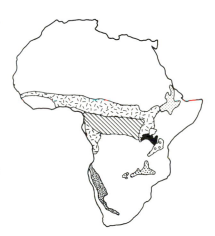

(hatched)	Agapornis swinderianus
(squiggle)	A. pullaria
(light dots)	A. taranta
(black)	A. fischeri (similar range to Rufous Tailed Weaver Histurgops ruficauda)
(fine dots)	A. personata
(dense dots)	A. lilianae
(stipple)	A. roseicollis (similar range to Sociable Weaver Philetairus socius)

All African lovebirds are closely related and have a common root with the Ring-necked Parakeet, *Psittacula krameri*, a species that can conveniently exemplify the lory stock that first entered Africa. It ranges from China to the Atlantic coast of Africa in a broad band that lies between the deserts and the forests. In spite of being a rather resident species, Ring-necked Parakeets are able to become very abundant and spread over wide areas because of their diet and the efficiency with which they process food (minute, hard but very nutritious seeds are instantly cracked by a small but powerful beak and manipulated by an exceptionally deft tongue). The diet of parakeets and lovebirds includes figs and simi-

The head of a Ring-necked Parakeet *Psittacula krameri* (left) and Collared Lovebird *Agapornis swinderianus* and the distribution of lovebirds in Africa (*above*).

lar fruits, flowers and grass seeds. It is the last item that has enabled parakeets and lovebirds to survive in very dry areas. Their major limitations are a year-round supply of food and a total dependence on the nests or nesting holes of other birds, crevices or naturally hollow branches to roost or nest in. A permanent food supply and dependence on hollows and other birds' nests only provide a partial explanation for the very patchy distribution of lovebirds today. A large part must lie hidden in climatic vicissitudes that fragmented the early lovebird population and led to the emergence of seven species, each with subtle nesting and food preferences. The two basic requirements of lovebirds would always have been present in the Zaïre basin under all climates. Today figs and flowers sustain Collared Lovebirds in the forest canopy, while the holes of woodpeckers and barbets probably leave them an abundant choice of nesting and roosting sites.

The Zaïre basin is surrounded by a ring of other lovebird species, all of which are predominantly grass seed eaters. To the north the closely related Red-headed Lovebird (*A. pullaria*) inhabits the open woodlands where the dominant trees are frequently hollow. A similar abundance of natural holes favours Lilian's Lovebird (*A. lilianae*) in the Mopane woodlands to the south. In Tanzania two lovebird species may be partially dependent on the nests of particular weaver species (see map). The lovebirds have evidently radiated from an original population that differed from modern parakeets in being very small (Collared Lovebirds weigh 40 g) and having a short tail. If the wide distribution of Ring-necked Parakeets is any guide the earliest lovebirds could have had a very extensive range in Africa. The Collared Lovebird's colouring is more like that of a parakeet than any other species and it probably represents the most conservative of the seven species. Oscillations of climate would have caused peripheral populations to drift and become stranded in outlying belts of favourable territory. Here their strictly resident habits and a limited choice of nesting and roosting sites would have encouraged speciation. If the Collared Lovebird is the least modified of contemporary species this could be due to long residence. The Zaïre basin would have suited lovebirds under all climates and from this viewpoint it can be seen as the stable geographic centre of their pan-African range.

It is clear the great changes of climate have forced many animals and plants to adapt or perish. Species wholly confined to forest or savannah have come and gone many times (possibly more than 20 times in less than 10 million years), but at each oscillation there could have been exchanges each way.

The contrast of extremes tends to overshadow a most significant category. For the animal or plant that is specifically *riverine* the Zaïre basin has remained habitable, wet or dry. The enormous variety of river types, smaller/larger, higher/lower, drier/wetter and their great extent, north and south, east and

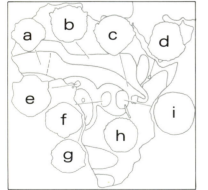

Opposite page, bottom. Some contemporary monkeys that are relatively terrestrial and might have derived from early expansions out of a dry Zaire basin. *a*, Green Monkey *Cercopithecus (aethiops) sebaeus*, *b*, Patas monkey *C. patas*, *c*, Tantalus Monkey *C. (aethiops) tantalus*, *d*, Bale Monkey *C. (aethiops) djamdjamensis*, *e*, Preuss' Mountain Monkey *C. l'hoesti preussi*, *f*, Salongo Monkey *C. salongo*, *g*, Talapoin Monkey *C. talapoin*, *h*, Vervet Monkey *C. (aethiops) pygerythrus*, *i*, L'hoest Mountain Monkey, *C. l'hoesti l'hoesti*.

Opposite page, top right. Hypothetical expansion of early populations while others remain as relict populations.

Opposite page, top left. Allen's Swamp Monkey, *Cercopithecus nigroviridis* with a map of the Zaire basin subdivided into five zones to show likely differentiation of early guenon population. The red area approximates to the present distribution of Allen's Swamp Monkey.

Overleaf. Congo Peacocks, *Afropavo congensis* (male above, female below).

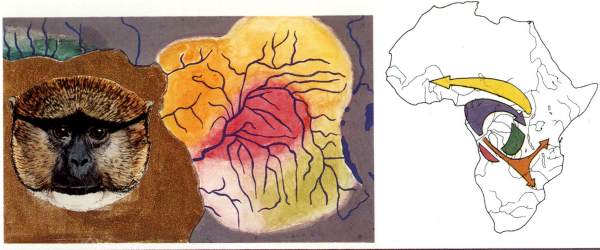

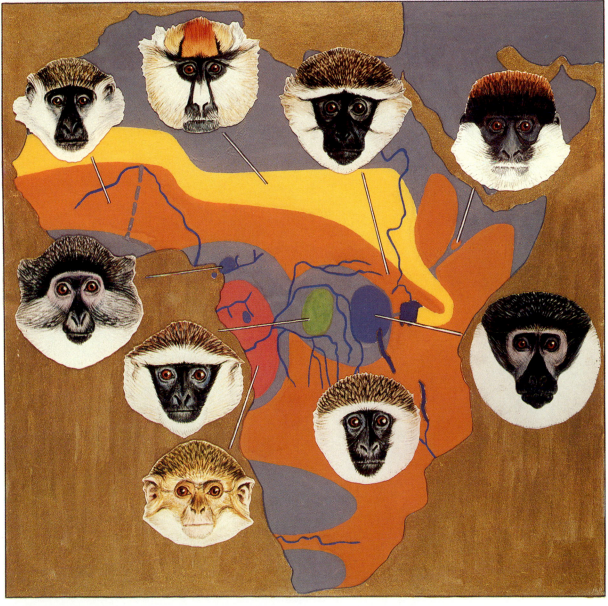

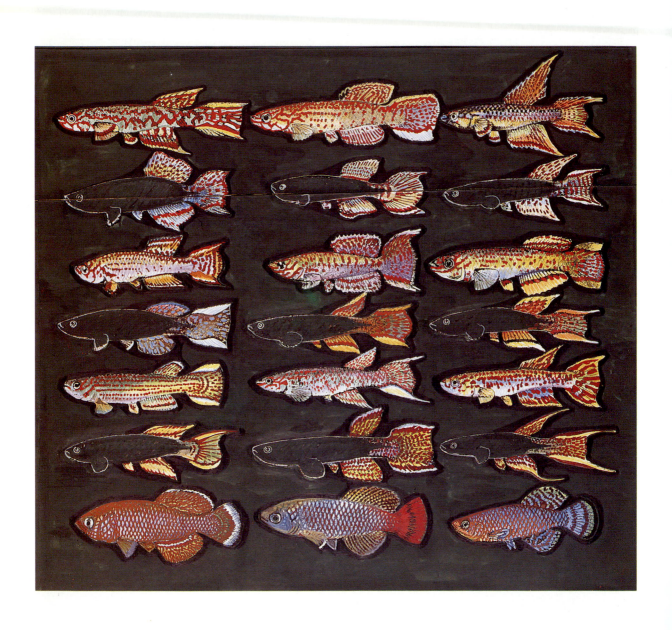

west, has ensured that there has been plenty of opportunity to differentiate. Speciation might have been easiest during dry periods, when physical isolation would have been more complete, but the basin's subdivision into distinct zones is equally real during wet periods. Speciation events may be obscured soon after they take place. For example, a population confined to a cool southern section of the basin during arid periods may later be the best fitted to use some well-shaded, cooler forest niche. With a return to moist conditions such a population can then expand and colonize the whole basin. Another species, from a region of dwarfed trees, may become adept at making a living in fine sunlit foliage. Released from its enclave by a change in climate, such a population may then colonize forest edges, windfall glades or even the canopy throughout the main forest block.

With time, a single group of organisms can proliferate in this way until the overlapping layers of species cannot be disentangled and their origins become as obscure as the waters in Luozi whirlpool. At the risk of stretching a metaphor, the Zaïre basin not only conceals a multiplicity of origins, it is an evolutionary whirlpool, a vortex where riverine organisms not only churn around within the confines of the basin but sometimes flow out to colonize lands beyond.

A prime example of speciation by a riverine animal concerns African guenon monkeys of the genus *Cercopithecus*, in which all but one of the 12 major lineages occur in the Zaïre basin. The simplest manifestation of the guenon radiation is a wide scatter of very localized species and subspecies. Studies of monkey chromosomes have shown that these belong to three of the more recently evolved species (their chromosomes number between 66 and 72, whereas the more conservative species range from 48 to 68). Graduations in the numbers of chromosomes and interpretation of their detailed structure has generated a family tree that corresponds very closely to the monkeys' anatomy and biology. When this tree is correlated with their distribution, the Zaïre basin emerges as the hub around which guenon evolution has taken place. A single evolutionary radiation of 12 lineages and some 27 species can, for the first time, be picked apart and the sequences (although not their timing) can be tentatively reconstructed, step by step.

Guenon distributions illustrate the consequences of climatic change. Local races occupy, in a rather consistent pattern, six or seven Centres of Endemism within the basin. But there is more to it than that. To retrace the guenon radiation from its beginnings is to follow a single biological 'model' through a series of minor but very significant modifications. The result?—Every type of wooded

Opposite page. Rivulin fish (males) with map showing limited distributions.
a. *Roloffia geryi*, Gambia to Sierra Leone. b. *Roloffia monrovia*, Liberia.
c. *Aphyosemion bivittatum*, Upper Guinea. d. *Roloffia gulare*, S.W. Nigeria.
e. *Aphyosemion sjoesti*, S.E. Nigeria/Cameroon. f. *A. bualanum*, Cameroon.
g. *Aphyosemion walkeri*, Ivory Coast/Ghana. h. *A. filamentosum*, W. Nigeria. i. *A. gardneri*, S. Nigeria/Cameroon. j. *A. mirabile*, Cameroon. k. *A. ahli*, E. Cameroon. l. *A. australe*, Gabon. m. *A. oeseri*, Fernando Po (Bioko) Island. n. *A. batesii*, Cameroon/Gabon. o. *A. louessense*, N.W. Zaire basin. p. *A. striatum*, N. Gabon. q. *A. cognatum*, Central Zaire basin. r. *A. christyi*, N.E. Zaire basin. s. *Nothobranchius brieni*, S.E. Zaire basin. t. *N. palmquisti*, Usambara region. u. *N. rachovi*, Maputo, S. Mozambique.

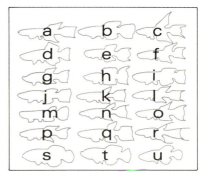

Previous page. Two species of Lake Malawi cichlid where sexes and 'moods' are colour-coded.
Top Pseudotropheus auratus, mature dominant male. *Below* female or subdominant male. Lower two, *Pseudotropheus tropheus*, male colouring blue, female colouring yellow.

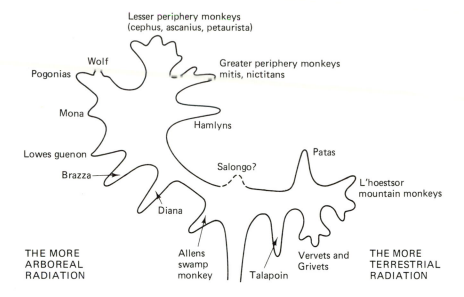

The guenon family tree.

habitat and sub-habitat in and out of the Zaïre basin has been colonized by a distinct form of guenon. The agency?—Climatic change operating on the unique geography of Africa's greatest river system to create a series of discrete 'islands', each with subtly different habitats.

In the short account that follows I have suppressed many of the alternative explanations and uncertainties that beset the real history of guenon evolution in favour of a simple bold outline. The simplification is deliberate. The objective has been to show that it is ultimately possible to understand the processes involved. None the less, the particular progression that is sketched out here is well founded in the realities of zoogeography, climatic history and the animals' biology.

So what is a guenon? *Cercopithecus* means monkey with a long tail, and this balancing organ implies arboreal habits, as do strong supple hands and feet and a relatively light weight (averaging 6 kg with a range of 0.6–25 kg). Unlike the large troops of long-faced terrestrial baboons, guenon groups usually consist of a small number of females, their offspring and a single male living in a stable home range. The exclusion of other males and the spacing out of groups is assisted by the dominant male's loud long-distance calls.

In a community of several monkey types, a more thorough use of resources is achieved by specialization in the type of forest, favoured heights and foliages, different foods and, possibly most important of all, in the techniques by which food is found and selected. A rich forest region may contain as many as six species and a single tree can attract up to four of them.

However, this assortment of attributes includes much that is of more recent acquisition, notably short faces and wholly arboreal habits. We know that the earliest guenons shared a common ancestry with baboons and macaques and that they were not always so different. Like baboons, the earliest guenons were largely terrestrial savannah animals, but they would have first emerged as a distinct, more slender, type of monkey because they abandoned the tightly integrated safety-in-numbers principle in favour of foraging more intensively in smaller, looser, but probably less mobile, groups. The incentive for this was provided by the pre-eminence of riverine environments during very prolonged arid phases. In dry country, fruit, gum, buds and insects tend to be found along drainage lines where there is more woody vegetation and moisture. Because long narrow riverine strips have strictly limited quantities of these typical primate foods they are better able to support a small number of small-sized resident monkeys than large numbers of large nomadic monkeys. They also provide better shelter for more vulnerable animals.

As a single river system that spans the greater part of tropical Africa the Zaïre and its innumerable tributaries must have been the main focus for this development. Even during the driest periods the cumulative extent of its tree-lined rivulets, streams and rivers would have been immense. Once colonized by the emergent archetypal guenon it could not have been long before some regional specialization appeared. For example, the trend towards smaller size could have accelerated in a region with numerous larger primates and a dense scatter of very fine-grain foods. By contrast, less competition, more predators and a wide scatter of food would favour larger animals. Likewise, monkeys in the Zaïre's upper reaches, both north and south, would have had very different conditions to contend with from those on the main rivers on the equator.

We can therefore look for an early separation of guenons into northern, central and southern populations (with a possible outlier in Upper Guinea), as suggested in the map on p. 204. There is no evidence for early guenon populations east of the Zaïre basin and it is conceivable that they were excluded from that area by baboons, mangabeys and colobus monkeys.

In the upper maps on p. 197 I have sketched out a scenario in which the northern population subdivides and expands into three regional types while the southern population splits into two. From such beginnings we can understand the present distribution of the five main 'terrestrial' guenons (shown in a somewhat schematic form in the lower map on p. 197).

A dry Zaïre basin was undoubtedly the source from which still drier outer regions were colonized; one of the earliest expansions was by a population that could be called the 'arid phase northern periphery monkey'. Its modern descendant is the Patas Monkey, *Cercopithecus patas*. This proto-patas was eventually displaced by the Green Monkey, *C. aethiops*, which, invading from the south, forced Patas stock to adapt to still drier, more open steppes along the southern margins of the Sahara.

Since this early radiation, some terrestrial monkeys may have moved very little, and somehow or other they have been able to cope with a long succession of alternations between forest and savannah. Presumably their primary attachment remains that of the archetypal riverine guenon. Well to the north-west and east of the main river are three highly localized forms of L'hoest's Monkey, *C. l'hoesti*, which are found only in moist forest. Another is one of the rarest of Zaïre basin endemics, the Salongo Monkey, *C. salongo*, which might represent the last remnant of an early 'arid phase southerner'.

Along the Atlantic seaboard and its immediate interior, a primitive stock of guenon monkeys became progressively smaller and more arboreal, a trend that culminated in that forest dwarf the Talapoin, *C. talapoin*.

It is known that the most conservative set of chromosomes (numbering 48) is possessed by Allen's Swamp Monkey, *C. nigroviridis*. This is a drab, khaki-coloured monkey that lives in palm and swamp forests on the *levées* of the Zaïre River. Genetically this monkey is separated from the West African Diana, *C. diana*, by some minor chromosomal rearrangements and five fissions (giving a count of 58 chromosomes). In spite of great differences in appearance between the two species, the discovery of this relationship makes it possible that a more truly arboreal lineage first emerged in the narrow coastal forests of Upper Guinea and that Dianas represent their modern descendants. With the development of a more arboreal type the stage was now set for the second major phase of guenon evolution.

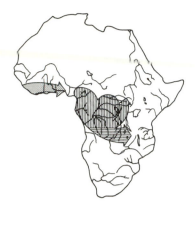

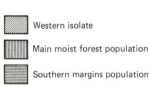

Western isolate

Main moist forest population

Southern margins population

Hypothetical diagram of earliest arboreal
guenon radiation during a wet period.

Subsequent isolation of populations
during arid periods with some
contemporary species that may derive
from such early differentiation, i.e. *C.
diana* (in far west), *C. neglectus* (Zaïre
basin), *C. hamlyni* (eastern Zaïre), *C.
mitis opistosticus* (south west of Lake
Tanganyika) and *C. campbelli lowei*
(western mona).

The arid phase 'terrestrial' lineages had already begun to diversify when rain
returned; the Zaïre basin became forested and therefore better suited to arboreal than to semi-terrestrial riverine guenons. Once such a monkey occupied all
of the now very wet Zaïre basin it soon began to subdivide along geographic and
ecological lines that were not dissimilar to those followed by its arid-phase predecessors (see p. 197). One or more dry interludes may then have intervened to
reinforce their genetic isolation. This fragmentation culminated in four 'wet
phase' types whose descendants are still recognizable today.

The very wet Biafran forests and Ogoue basin were dominated by a medium-
sized type which eventually expanded westwards (where the modern representative of these proto-mona monkeys is *C. campbelli lowei*).

A second species colonized the swampy centre. *Cercopithecus nigroviridis*, the early
'dry phase swamp monkey' was not wholly supplanted, but the very successful
Brazza Monkey, *C. neglectus*, is descended from this stock of 'wet phase swamp
monkeys'. In the drier, cooler forests of the south an arboreal monkey developed
that is ancestral to so many species that it is best designated as the 'wet-phase
periphery guenon'. An intermediate type between the last two had a range that
may have been partly interposed between them. Its modern descendant, the Owl
Monkey, *C. hamlyni*, is one of the least-known endemics of eastern Zaïre.

The schematic map above shows the four wet-phase populations (with portraits of four contemporary descendants).

Proto-monas and proto-brazzas, being of the same stock, would have shared
the archetypal guenon predilection for riverbanks. (west African monas still
show some preference for them.) However, western Africa and the central

Zaïre basin have always been very different habitats and the two monkeys'
divergence would have been further accentuated by contrasting ecological
strategies. Even today the heavy Brazza is a slow thorough forager that has not
entirely abandoned ground feeding. The monas instead combine more arboreal
habits with smaller size and greater agility. These might have been mere trends
at the time the two populations were first separated, but when the proto-monas
eventually invaded the northern Zaïre basin they would have become very specific differences.

As the proto-monas expanded eastwards they entered primate communities
where swamp monkeys and proto-brazzas were not the only monkeys to exclude

them from living along the riversides. Eastern monas were therefore subject to strong selection for smaller size because there was less competition for food and living space among the smaller branches and finer foliage. They therefore came to occupy a very different niche from their western counterparts, and became the first wholly arboreal canopy guenon. Today the eastern and western monas retain very similar coat patterns although their niche and body size have diverged quite significantly; because of this similarity both were formerly lumped together as mere races of a single species. Now it is known that

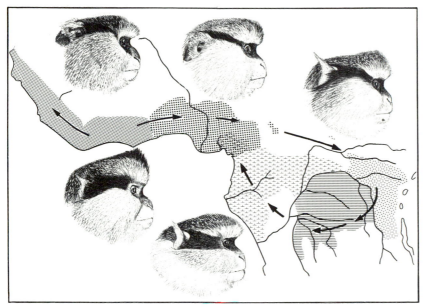

The radiation of mona monkeys interpreted as expansion by an early form of wholly arboreal monkey out of a western focus. The overlapping ranges of two mona monkeys in Cameroon is interpreted as the product of recent westward expansion by the small *C. pogonias*, possibly at the expense of the larger *C. mona mona*. It is possible that the range of the latter may be slowly contracting.
Heads. *Top left, C. campbelli lowei, top centre, C. mona mona, top right, C. denti. Bottom right, C. wolfi, bottom left, C. pogonias.*
Distribution. Fine stipple, *C. campbelli*; heavy stipple, *C. mona*; open dots, *C. denti*; hatching, *C. wolfi*; inverted T, *C. pogonias*.

chromosome counts differ for several monkeys in the mona complex, it is regarded as a superspecies which contains at least three full species and possibly as many as six.

The eastern Dent's Guenon, *C. denti*, is probably much closer in its ecology to the distinctively patterned Crowned Guenon, *C. pogonias*, which occupies all the forest between the Zaïre River and Cameroon. Between them is an intermediate form, Wolf's Guenon, *C. wolfi*, which lives south of the river.

The mona monkeys therefore form a ring around the heart of the Zaïre basin. The smallest and most specialized, the Crowned Guenon, seems to have expanded very recently, probably at the expense of the older *C. mona* (which gives every indication of being in the process of replacement by the Crowned Guenon).

If you imagine the whole process speeded up, the monas become a wave which sweeps east, then down, round and back so that the crest breaks over its ebbing backwash. The monas have made a great circular sweep through the Zaïre basin.

Overleaf. Hamlyn's Owl Monkey *Cercopithecus hamlyni.* A sombre black and grey monkey with vivid cerulean blue patches on the thighs in both sexes and blue testicles on the male.

setting
erect changes
very common + displays
blue v. effectively

Exhibited to
grown + male
hanging upside down

seated to show
blue thighs

squat on
in back
view

hamlyns threatens in
low forward chin thrust
with slight brow
movement only
yawn. Does sideslams but a
very sticky up + down hop jump in
mock threat or general excitement

Swivelling of this
head mask is conspicuous
& a lop-sided twist was
given at me by this ♂
+ head movements are
continuous

cerulean
blue area

infant
yellow
underparts
greenish brown
grizzled back

They are one peculiarly symmetrical wave in an evolutionary whirlpool. It is a whirlpool that may well have churned up innumerable other species.

Now slow the same process down and consider the pauses. Each cell of the basin holds a distinct mona population, which suggests that the monas changed at each pause. In most cases big rivers mark the boundaries of each population and these long stretches of water have evidently helped to keep closely related populations apart.

Refuge areas in and around the Zaïre basin. Boundaries of contemporary forest zone indicated by broken line.

It is less likely that these rivers were the only barriers that originally allowed populations to become distinct. The succession of dry periods that is known to have affected Africa since the Pliocene is likely to have been the prime mechanism for isolation. Each episode would have been long enough and the separation thorough enough, for local adaptations to become pronounced. The distribution of monkeys (and some other organisms) suggests that there were three major lowland forest refuges in the west, east and centre of the main forest block, and that these subdivided into a total of 9 or 10 lesser refuges. The Niger delta, Cameroon and Ogoue basin were the three most westerly foci. The confluence of rivers and lakes in the centre of the Zaïre basin was the site for a very distinctive swamp forest refuge, while the 'Ituri–Maniema forest refuge' covered the moist eastern margins of the basin. The two major tributaries of the Zaïre River, the Lualaba and Lomani, enclose a long narrow strip of low hills, isolating populations in a minor 'fluvial refuge'. The fauna here is not always distinct from that of the southern scarps of the basin. Rain from the Atlantic seems to have maintained narrow forest refuges near the coast of Angola.

That each of these refuges differs in elevation, soil and climate is already known. It is also known that each has its own community (of which predators, prey and local diseases are an integral part), but for any one invading species it is closely related predecessors or competitors that are most important. One or more of its fellow-primates or fellow-guenons will be the most potent influence on what the new arrival can or cannot do. Within such enlarging communities each species must adjust, and in such adjustments lies the course of evolution.

Ultimately, resources determine how many species an area can support. Richer areas have more, poorer areas fewer and the monas have only succeeded in the richest lowland rain forests because they are excluded from the more peripheral forests.

The guenons that put such limits on the monas are close relatives. They derived from the very first division of wet-phase monkeys. Return then to those early beginnings to trace the fortunes of these, the most diverse and widespread of African forest guenons. From their first enclave (probably on the south-eastern margins of the Zaïre basin) the 'periphery guenon' soon expanded around

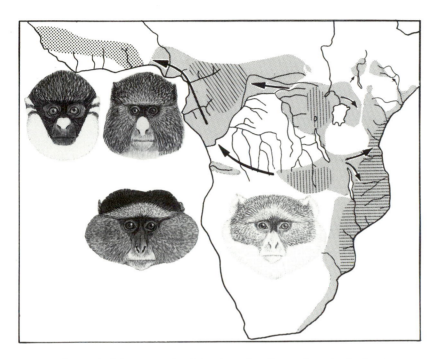

Suggested spread of the greater 'periphery' monkey from a focus in the South-eastern Zaire basin. This expansion may have given rise to 4 major populations: the *mitis* group in Central Africa (vertical hatching); the *albogularis* group on the eastern littoral; *C. nictitans* in Cameroon Gaboon and a smaller western isolate that subsequently became the *petaurista/cephus* group.

the margins of the main forest block and thus it eventually arrived in Upper Guinea. Here it could have encountered more than one guenon, but the dominant species would have been the large 'early arboreal' or proto-Diana.

With contemporary Diana as a guide it can be predicted that the new arrival would have been excluded from high forest. Not only would its niche have been constricted by pressure from one or more competitors, but its preferred habitat would have lain sandwiched between the high forest and drier savannahs inland. Proximity to the Sahara meant that this belt was both exceptionally narrow and vulnerable to degradation.

Competition and a confined unstable niche could help explain why these monkeys became smaller in size in Upper Guinea. They would also have gained the little monkey's basic advantages of small but nutritious foods out on the thinner branches of secondary growth. Elsewhere, size was less of a constraint and the 'wet-phase periphery guenon' developed such a wide ecological tolerance that it soon expanded its range out of the Zaïre basin.

The Tanzanian highlands were a bridge to the entire length of Indian Ocean littoral. Another wave swept up the Rift Valley to the Zaïre–Nile watershed.

Eventually they even reached Ethiopia (from two directions), but their expansion was probably rather erratic over a very long period of fluctuating climates. In Cameroon, their modern representative is the drably coloured Putty-nosed Guenon, *C. nictitans*. Today more than 20 regional races are recognized as belonging to the *mitis/nictitans* group. These 'greater periphery guenons' have remained essentially marginal in distribution and are largely absent from much of the central forest block.

By contrast, their cousins the dwarfed or 'lesser periphery guenons' eventually broke out of their Upper Guinea enclave and entered the central forest block. Although they spread very widely through its outer forests, the monkeys also penetrated right to the centre of the basin. This can be explained by the fact that secondary growth springs up within mature forest wherever there are windfalls, landslides, drownings, fire, flood or elephant damage, all of which generate good habitat for the 'lesser periphery guenon'.

Modern monkeys of this type are classified as the cephus superspecies, *C.* (*cephus*). The group embraces 6 regional species and 13 subspecies, 10 of which occur east of the Niger. Once again the now familiar pattern of subdivision within the main forest block can be invoked to explain speciation in the cephus group, and their exceptional diversity is clearly the product of more than one fluctuation in the rain forests. Four contemporary populations are confined to small localities, where they give every appearance of being slowly 'squeezed out' by more expansive neighbours.

Throughout their range these small monkeys are almost identical in ecological niche, physique and behaviour. Yet the six species have diverged in one particular respect: their brightly coloured or 'contrasty' face patterns are as unlike each other as any one of them differs from monkeys in other lineages. The explanation for this anomaly lies in the development of a sort of signal code whereby friendly intentions and 'greetings' are signified by very fast head movements. At the time of their original dispersal through the main forest block, cephus monkeys must have had relatively drab faces which were very poor transmitters for their unique head-flagging messages. Speckled furry cheeks and cap, a white chin and faintly tinted bare face-skin—unpromising material with which to weave a head-flag, yet the cephus monkeys' situation made their face an obvious site for a small but highly conspicuous signal, a signal they really needed. In peripheral habitats monkeys have a relatively 'quiet' environment—when they enter the enlarged primate communities of the main forest, life gets very noisy. Apart from voices and the bashing of foliage, crackling of branches etc., there is *visual* noise to compete with. All the movement and jostle of birds and other animals takes place in a sun-spattered chaos of tropical vegetation where shiny leaves reflect sunlight through a matrix of dense velvety shadows. The postures, patterns and grimaces of six or seven species of monkeys generate still more visual noise. Amidst this crowded marketplace the very visual cephus monkeys transmit their own messages in their own codes with nods, shakes and weavings of their heads.

Their faces illustrate what a variety of different patterns can be elaborated from quite simple rearrangements of tone and colour with the odd pleat, whorl or cowlick of the fur to give a little sculpture to the mask. Some rely on colour contrast—blue versus yellow, red against cream, black and white. Then there is geometry—horizontal, linear, circular, 'target' patterns or an accentric V or S.

What is especially interesting about the head-flagging is that it also comes into play during courtship. The patterns might therefore help to keep the species distinct as well as to maintain separate populations of cephus monkeys. In fact hybrids do occur in the wild but in spite of this ability and occasional readiness to cross-breed, each guenon species gives every sign of being just that—a species, a distinct entity.

In this condensed account I have described a succession of closely related populations spreading through Africa's forests, layer over layer. Each new invader and each of the invaded has been presented as a distinct adaptive type that has maintained its identity. So, to all practical purposes, they have; but recent chromosome studies, using new techniques, have revealed that, in spite of all appearances, many if not most guenon species are effectively hybrids. Even as they present all outward signs of more than 20 'good' species, guenons have *not* remained genetically pure and intact.

This unexpected discovery confirms that entire communities, species and now genes have been churned in this evolutionary whirlpool at the centre of Africa. A new and dramatic theatre for science and natural history is only just being revealed.

Colobus monkeys and mangabeys present a similar but more irregular patchwork of populations to those of the guenons, but both belong to radiations that preceded and perhaps anticipated that of the guenons. The contemporary species are mostly specialists in peculiar niches with peculiar diets or ecological strategies. It is quite likely that many earlier types were displaced by guenons, but there is no direct evidence for this.

Apes, older still in the primate tree, are now only represented by three African species. One of these, the Pigmy Chimpanzee, *Pan paniscus*, is strictly limited to the swampy inner Zaïre basin, south of the river. Here levées and sump-lands probably permitted this great ape to survive several arid periods. As with so many Zaïre species, the Pigmy Chimp is probably more versatile than its present very limited range and habitat suggest.

Earlier in this account the biogeographer's title 'Ituri–Maniema forest refuge' was used for the moist eastern margin of the Zaïre basin. Although a part of the basin, Ituri–Maniema is comparable with that western focus, Cameroon and Gabon, as the main central African focus for lowland forest life.

The Okapi is not the only unusual animal to occur in this region, the Fishing and Giant Genets and Congo Peacock are among the more spectacular species from this refuge area. The Fishing Genet, *Osbornictis piscivora*, is a relative of the genets and palm civets that has adapted to catching fish and frogs in streams

Fishing Genet, *Osbornictis piscivora*.

within the forest. As with other genets it is clawed like a cat but has naked toes and a short bewhiskered nose somewhat like an otter. It seems to be a surface fisher not adapted to pursue its prey under water, as otters and otter-shrews are, nor to feel for them in murky waters, as the Marsh Mongoose does. As small fish-eating mammals there could be some marginal overlap between all these species but a clue to the Fishing Genet's very restricted range lies less with mammalian competitors than with birds. The majority of fish-eating birds, whether herons, storks, kingfishers or fish-eagles, hunt by sight, plunging down on their prey from above. The Fishing Genet seems to suffer similar limitations and therefore can only compete where such birds are at a disadvantage, notably where shallow clear streams are densely obstructed and overgrown.

Such small forest streams occur in many parts of Africa today but they are an early casualty of drought. Drying up or opening up, the periodic fate of such streams elsewhere would have been fatal events for the Fishing Genet. Not only is this animal likely to be a gauge for stable climate and environment in the region, but also the conditions there are probably close to those in which it first evolved.

The Giant Genet, *Genetta victoriae*, is also aberrant. Most genets range between 1 and 3 kg in weight and tend to be very arboreal. About twice the size of the Fishing Genet and a predominantly terrestrial animal, the Giant Genet feeds on fruit, insects and small mammals. This is a similar range of foods to those taken by the still larger and more robust African Civet; none the less the rare genet and very widespread civet co-exist in this section of the basin. A proliferation of small carnivore niches could be taken to indicate the overall richness of the forest, but it is the long-term climatic stability of the region that has probably been decisive. Sandwiched in between the well-tested genet and civet niches, the Giant Genet may have avoided being 'squeezed out' by its relatives because the region has not suffered the great changes in climate that have affected most of the rest of Africa.

Stability may also have contributed to the survival of the Congo Peacock, *Afropavo congensis*, in eastern Zaïre. These magnificent birds (iridescent blue, green and black in the male; tiger-striped in Indian red and green in the female) share the forest floor with three francolins and three guineafowls, all of which are much smaller species. Named and described for the first time in 1938, the Congo Peacock is presumed to be the relict of a widespread early stock of peacocks. These were unlikely to have been true forest birds, especially as forest connections with India were broken very early on. Although the contemporary species is found under a dense understorey it prefers fairly dry, often secondary or 'edge' forest, which is consistent with riverine galleries having been the most stable element during Zaïre's fluctuating climatic history. Like the Oriental Peacock, the male spreads its short tail in courtship and bows to the female. In the morning and evening they utter a very loud cry. Once identified, these calls revealed that the birds were scarce, well spaced out (but very unevenly) and that they wandered widely. They soon disappear from settled parts of the forest. Of several other birds with Oriental origins, that primitive woodpecker the piculet, *Sasia africana*, is, like the peacock, peculiar in its isolation. Here too the minimal stability provided by riverine forest may have helped survival.

As a single vast river network the Zaïre basin is unlike any other region of Africa. The river dominates animal and human life alike and gives its name to the Republic of Zaïre, which covers the greater part of the basin. The 4 million sq km of the basin currently support some 40 million people. Zaïre has some of the finest national parks in the world, but these are mostly in the far eastern side of the country. One, the Salonga National Park, covers a significant area of the central basin. Another, Maiko, is in the north-east. It is hoped that other reserves will eventually be upgraded so that each basin 'cell' and its unique flora and fauna will be adequately represented.

Western scientific interest in the Zaïre region found its first formal expression at the Brussels International Exposition in 1897, at which time the Congo Free State had been the personal empire of King Leopold of the Belgians for a dozen years. A Congo museum was set up after the Exposition closed and material for it was gathered by officers of Leopold's state. For the next 20 years or so many expeditions were little more than hunting excursions by European aristocrats such as Duke Adolf of Mecklenburg, Prince Wilhelm of Sweden and Prince Leopold. On Christian mission stations individual missionaries such as Fathers Goosens, Callewaert, Longo and Van Assche became keen naturalists, and natural history has remained very much the province of private enthusiasm and initiative among residents and visitors alike.

The discovery of the Okapi in 1900 focused the biological world's attention on the Congo and led to what has remained the single most significant work of zoology to be conducted in the Congo. In 1909 the American Museum of Natural History appointed Herbert Lang and his assistant James Chapin to conduct a comprehensive six-year zoological survey. The results of their expedition and those that followed in 1926 and 1931 were still being published up to 1954.

With the formation of national parks in the 1920s, research began to acquire specific bases and foci for sustained study. At about this time the *Bulletin du Cercle Zoologique Congolais* provided a medium for dissemination of the results of natural history research.

In the post-colonial era, Zaïre's closest ties with the West have remained with Belgium. Under M. Aniset the national parks kept up a practical working relationship with scientific organizations such as the Belgian universities and the Royal Museum at Tervuren and outstanding individuals such as Jacques Verschuren. Under the auspices of Kisangani University and the Zaïre Institute for Nature Conservation, biological exploration has continued with expeditions such as those of Antwerp and other Belgian universities. Kinshasa University Zoology Department has a working relationship with Liège University.

The knowledge of local people in Zaïre is still inadequately recognized. More than 100 years ago Harry Johnston noted that some Zaïrois had at least seven terms to describe different kinds and conditions of forest. The potential for creative partnerships between Zaïrois and foreigners remains largely unfulfilled.

In a village near Kwamouth in 1883 Johnston pronounced the houses well made and well kept, the people prosperous, and things orderly, material and thoroughly positive,

'There was one child I shall always remember with affection . . . she constituted herself my little guide, taking my hand with the greatest confidence and leading me through the village to show me the sights. Seeing me gather flowers to preserve she afterwards presented me with an armful which she had laboriously plucked and later she pressed into my hand 3 newlaid eggs, warm from the nest.'

As if he had a foreboding Johnston wrote: 'They tell me I am the first of my colour who has set foot in their village . . . May they never look back to it with sorrow, as marking the advent of a new and troublous change in their hitherto peaceful annals.'

The people around Kwamouth ferry are still friendly, material and thoroughly positive, but the Western world has been slow to recognize the extraordinary interest and beauty of Africa's greatest river, its people and its animals and plants.

CHAPTER 11

Lakes and Rivers

'What are you digging for?' I asked the two youths as they cut and levered away with panga and sharpened stick at a tangled root in the reed bed: 'Mamba.' A crocodile? I recoiled, but it was no crocodile that emerged from their spading and probing. Tearing away another clod, one youth plunged his hand into the cavity beneath and drew out a pale sausage folded over itself and covered in thick messy slime. The indeterminate creature writhed in slow motion as it was skewered behind its blunt head and I too shuddered as I watched the young men set off with it towards Mwanza town.

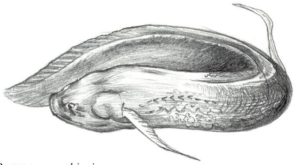

Lungfish, *Protopterus aethiopicus*

Animals are invariably enlarged in a child's memory, so I cannot now say what the real size of that lung-fish was. Brief as it was, the encounter still retains a trace of the special mystery that belongs to the unexpected and unexplained. Later, initiated into another sort of excavation—fossil-hunting—the incident acquired a new significance and context when I learnt of 'The Age of the Lung-fish' more than 350 million years ago. There were droughts, I read, in the Devonian and those early freshwater fishes were likely to have developed lungs to cope with them.

The African Lung-fish, *Protopterus aethiopicus*, is just one of many fishes that have found and refined solutions to life on a drought-prone continent, but there are huge differences in time-scale. The lung-fishes, together with their distant but equally ancient relatives the bichirs, *Polypterus*, go back to a time well before the break-up of Gondwanaland. Lung-fish have never become marine, so their presence in Australia and South America as well as Africa is a real measure of their antiquity.

Bony fishes developed much later, in the Jurassic, but well before Africa's connection with South America was broken. Among the many primitive tooth carps or cyprinodonts that occur on both continents are 'annuals' or rivulin fishes which are even more remarkable than the lung-fishes in their adaptations to drought. Rivulins are represented in Africa by four genera of very small colourful fish which are well known to tropical aquarists as killifishes or lyre-tails. Certain species of African rivulin can remain in aestivation for more than five years, but unlike the lung-fish they aestivate as an egg. The development of that egg is unlike any other known. After fertilization the cells separate into two populations that disperse within the egg's membrane. One type later joins together again to become the embryo. The other blastomeres become a blanket of cells that wrap around the embryo and can be damaged or destroyed without affecting the embryo, of which they become no permanent part. These extra-ordinary developments take place inside an egg which is buried in the mud, where its tough outer membrane can resist water loss over one or more dry seasons.

The trigger that breaks the embryo's resting stage or 'diapause' is sustained rain: the egg hatches within days and the fish grows and matures extremely rapidly—one species has been recorded spawning within four weeks of hatching. To grow and mature over such a short period is only possible for a very small animal and the tiny size of rivulins allows them to colonize the smallest and most temporary of wet-season pools, and because their survival depends on eggs, not free-swimming adults, they are not dependent on permanent water. Although many are destined to survive no more than a few months, captives of more than one species are known to live for five years. Their chances of individual survival vary greatly from region to region and species to species.

The most widespread are false-gills, *Nothobranchius*, which range from arid Somalia and northern Nigeria to southern Africa. Further west is another genus, *Roloffia*, which prefers moister habitats. Their close relatives, the killifishes, *Aphyosemion*, are restricted to shallow ephemeral waters within true rain forest. Like many other arid-adapted organisms they have found a niche within the forest which was closed to more conventional types of fish. Flash floods, elephants and meandering swampy rivulets all create temporary ponds under a dense canopy from which drop ants, spiders and insect larvae to feed the fish.

The speed with which these fish develop is matched by the intensity with which they pursue reproduction. Females lay between 3 and 50 eggs a day, but only 1 or 2 at a time. As a result, fertilization must be sustained almost continuously and there is much competition among males to repel or subjugate other males. Females are attracted to the winners. The social consequences are variable: monogamous pairs, harems, promiscuous hordes and male hierarchies have all been observed. Because ponds are frequently muddy or deeply shaded the fishes need strong cues for sex. Dominant males attract the mature females (which are generally drably coloured) by exuding chemical signals and making colourful displays; some, like chameleons, can change their body colours. In some species the variability of male colours and patterns is very great. For example, A. Brosset, one-time director of IRET in Gabon, reared three male offspring from a single pair of Bates' Killifish, *A. batesi*, that differed as much from each other as many recognized species do.

It is therefore possible that a few of the 30 or so killifish, 20 false-gills and other known rivulins are not full species; indeed, many varieties hybridize in aquaria. Nevertheless, it is certain that many very well-defined and distinctive species of *Aphyosemion* are highly localized, rain forest species endemic to the classic terrestrial refuges of Upper Guinea, the Bight of Biafra and the Zaïre basin. Some of the false-gills likewise seem to be restricted to limited parts of the east coast littoral, Usambaras and tributaries of the upper Zaïre.

It is the killifishes' fin patterns, bright body colours and elaborate displays that have seduced aquarists (the practical convenience of being able to send their eggs by post and keep the adults in bowls have been other attractions). Male rivulins display at each other with shudders and sudden jerks of heads and fins. They try to displace one another by violent blows of the tail but seldom resort to bites. Courtship is also accompanied by fin jerks and much shuddering before the male comes alongside and enfolds the female with his colourful fins while he fertilizes her eggs. Perhaps she shares in the submission of the dominated, but she is also drawn by the size, colour and scent of the male. She finally capitulates to his touch by expelling her egg.

In spite of their great variety, both within and between species, the patterns and colours of fins and tails follow certain design principles that are interesting in themselves. In the males, tails and fins are enlarged by extending struts along the outer margins of the fin, its centre, or asymmetrically. When erect, these extensions substantially enlarge the size of a male. In order for this enlargement to have its maximum visual impact in a dark or muddy setting, the outer or leading edges of fins and tail have a margin, band or bands of very light or bright colours. Tail and fin patterns sometimes resemble each other (in a few cases they even give the impression of a single scarcely interrupted plane), or they are emphatically different in colour, shape and design. Wherever anything is known of their behaviour these differences correlate with quite specific fin or tail displays.

Each male pattern has recognizable resemblances with the immature pattern that preceded it and with that of the hen fish. The latter have modest contrasts in light and dark, and subtle nuances of warm and cool colour. In males these become brilliant contrasts of ruby and azure, sapphire and emerald green. A filigree of female or baby browns becomes an iridescent mottle of spots and lines which tend to polarize on flags of fin and tail but also form a carnival mask around the black and silver eyes. Male killifish become the parading pheasants, peacocks or strutting cockerels of the fish world.

The development of such an extraordinary life-history epitomizes the ephemeral nature of water on a dry continent. It is no accident that lung-fish, bichirs, rivulins and the aestivating cat-fish are all archaic lineages—these are the precursors of modern fish and also among the most successful inheritors of Africa's scarce waters.

Being independent of permanent water, however, rivulins are quite unlike other fish, which are generally tied to particular river or lake basins. In fact, rivulins are rarely found in free-flowing rivers or lakes and a history of the fishes found there is very largely a different story.

As Africa's land surfaces have arched up or tilted into mountain massifs, split open as rifts and erupted in volcanoes, water has met new obstacles, followed new inclines, collected in new basins and carved new valleys. Some basins, notably Zaïre, have had their headwaters radically remodelled but remain very ancient. The existence of a lung-fish, *Protopterus dolloi*, and two bichirs provide one measure of the Zaïre basin's age; another lies in nearly 1000 fish species belonging to many families and adaptive types. Fishes, as the premier inhabitants of Africa's rivers and lakes, reflect the continent's geological vicissitudes in a way terrestrial animals cannot. Many are so tied to a single river system that a watershed between streams 1 km apart can mark out a boundary as momentous as the sea between two islands.

Mwanza, the scene of my earliest encounter with the lung-fish and 'living prehistory', is close to just such a watershed. The little town is perched on the southern shores of Lake Victoria, reservoir to the Nile. Less than 100 km southwest, behind the lowly Siga hills, lie the first streams of the Malagarasi, easternmost headwaters of the Zaïre basin. About the same distance south-east sees the beginnings of the Eyassi drainage, now enclosed within the eastern Rift Valley but once falling away to the Indian Ocean.

The three basins have very different fauna. The waters of the Malagarasi are interrupted by the 650 km trench of Lake Tanganyika before they join the Zaïre. But this swampy river supports several Zaïre riverine fish which are thought to pre-date the 3–6 million-year-old Tanganyika rift. Among them are two *Polypterus* species (*P. congicus* and *P. ornatipinnus*) which in spite of their Palaeozoic ancestry have remained endemic to the Zaïre river system.

There are more than 20 other river systems, each with its own specialities. They are strung around the coasts of Africa, but the largest rivers cluster in

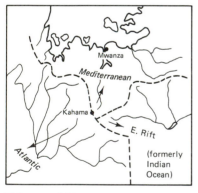

Kahama and Mwanza are close to the great divide.

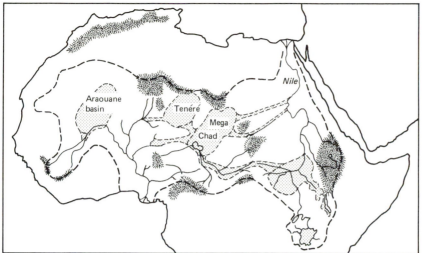

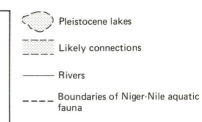

Pleistocene lakes

Likely connections

Rivers

Boundaries of Niger-Nile aquatic
fauna

Elevated areas of watersheds

The evidence for wet climates in Africa with Late Pleistocene shallow lakes in the southern Sahara and the likely connections between the Niger, Chad and Nile basins (modified from maps by Beadle, 1900).

three major groupings to produce, on the face of it, a rather surprising pattern. In geographic area if not in biological diversity, the Zaïre is almost dwarfed by the Nile–Chad–Niger systems which have exchanged much of their fauna relatively recently. The interconnection was along a series of shallow lakes across the southern Sahara, and the amalgam that has resulted is known as the 'Soudanian aquatic fauna'. A third system, the Zambezi, flows east into the Indian Ocean.

A major distinction between these three faunal regions and the river basins outside is that each great river is associated with a large lake. Lake Tanganyika flows into the Zaïre, Lake Victoria into the Nile and Lake Malawi into the Zambezi. In each case the duration of that association has been very different. Lake Tanganyika, of uncertain date but at least 3 million years old, is twice as old as Lake Malawi, while Lake Victoria in turn is half that age, a mere three-quarters of a million years.

The three lakes also differ greatly in depth and in the manner of their formation. Lake Tanganyika filled a rift that cut abruptly across the west-flowing rivers that preceded it, and at nearly 1500 m is only exceeded by Lake Baikal in depth. Lake Malawi is about half as deep and fills a 600 km trench that follows the line of an older tributary to the proto-Zambezi. Lake Victoria is shallow, nowhere deeper than 100 m, yet it covers an area about the size of Ireland or the state of Maine.

Shells from Lake Tanganyika. *Top row from left,* Spekia zonata, Chytra kirkii, Paramelania damoni, Tanganyicia rufofilosa. *Lower row,* Neothaumia tanganyicense, Bythoceras iridescens, Tiphobia horei.

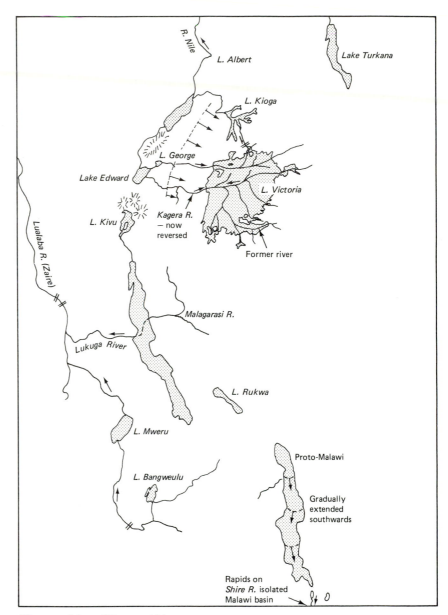

The great lakes of Eastern Africa.
Note the following historical events.
1. *Lake Tanganyika*. Approximately 3–6 million years old, possibly older. Former water level lower. Now very deep. Malagarasi River, a remnant of ancient Zaire river system cut by Lake Tanganyika. Pleistocene eruption of Virunga volcanoes dammed rift valley to form Lake Kivu which overflowed into Lake Tanganyika, where raised levels led to overflow into Zaire basin via Lukuga River.
2. *Lake Malawi*. Approximately 2 million years old. Earlier levels lower and lake formed by staged southwards extensions. Now very deep, still deeper in late Pleistocene.
3. *Lake Victoria*. Shallow, formed by Pleistocene uplift in west blocking westward flowing rivers (thus reversing flow of Kagera). Eventually overflows into old valley systems in north to form Victoria Nile and Lake Kioga.
   During very wet periods Lake Turkana (Rudolph) emptied into flood plain of the Nile above the Sudd.

Lake Tanganyika had access to the oldest, most varied river basin fauna, it is the oldest lake and it is the most radically different from a river, indeed, more like a sea. For these reasons its fauna has evolved furthest. When John Hanning Speke arrived on the shores of Lake Tanganyika in 1856 he picked up some of the mollusc shells lying on the shore. One that was named after him, *Spekia*, is remarkably like a marine cockle; compact, thick and rounded. Another like a *Cardium* sea shell, *Grandideria burtoni*, was named after Sir Richard B. Burton, his companion. Many others resembled sea shells very closely indeed, especially those with ornamented whorls, spiny projections and protruding siphons. Molluscs from muddy bottoms have spines to keep them from sinking into the anoxic poisonous sludge below, while thick-shelled smoother types live on rocky shores where they can withstand pounding by waves.

In 1883, a coelenterate, *Limnocnida*, which closely resembles a marine jelly-fish, was found in Lake Tanganyika. Together with the 'sea shells' and the presence of two sardines belonging to the herring family, this was seen as evidence for a direct link with the ocean. The theory of a Jurassic sea connection lasted until 1920, when a comparative study of east African lake faunas by Dr W. A. Cunnington showed that all the 'marine' species had recognizable freshwater relations and that their peculiarities had been evolved during their long isolation sometimes in water as salty as seawater.

Lake Malawi is similar to Lake Tanganyika in size and conformation but not in depth or age. It is also enclosed in a much younger, smaller and more impoverished basin and so has had a much smaller reservoir of colonist species. As a consequence its contemporary fauna, although numerous, derives from a very small number of founders.

Drowned inselbergs around Lolui Island, Lake Victoria.

Lake Victoria's present appearance is most misleading. It is not only young but has had a much less stable history than the older, deeper Rift lakes. Lake Victoria began its existence when the westward flow of two large rivers was blocked by the lifting of a long ridge along the eastern margins of the western Rift. These rivers had originally been tributaries of the Zaïre, but their middle courses had later been diverted to join the Nile. As their waters ponded back, a mosaic of small lakes and oxbows would have formed. Their enlargement may have been intermittent but the basin eventually filled until it reached its peak about 30 m above the present level. This did not last and an arid period about 15,000 years ago reduced the lake to a ragged network of closed saline puddles. Such vicissitudes might have occurred more than once, driving the fauna between rivers and lake, small lakes, large lakes, fresh water, salty water. With

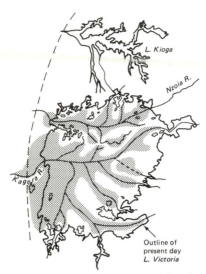

The river system now submerged under Lake Victoria.

each fluctuation there were opportunities to speciate and proliferate. Today there are over 200 endemic fish, of which three-quarters belong to a single type known as haplochromines. It is possible that they derive from one (at most four) riverine ancestors.

These are cichlid fish and their separate and independent development in each of the three great lake systems rivals the fauna of the Galápagos as an example of evolution. From different sets of ancestors the cichlids have radiated into a similar range of types in spite of each lake having very different origins, ages, depths and ecologies. There are numerous parallels or convergences between the three cichlid faunas but they are also graded. The most extreme and evolved forms come from Tanganyika, the least so from Victoria.

Why has the same type of fish been recruited to populate the three lakes? There are several reasons: the lakes originally formed close to headwaters that were thousands of river kilometres from the sea, and the choice of colonizing types was therefore quite small. Because they tend to be ephemeral, inland waters demand from their inhabitants specializations that are not easily modified or put into reverse. Lung-fish, bichirs and rivulins have strategies to overcome drought or anoxia but other less specialized types also have to solve oxygen problems. Many need to gulp air from time to time if only to regulate or prime their swim-bladders, while the vast majority of river fish depend on flowing water being fully aerated for their own needs or, even more crucially, to hatch their eggs.

To physiological limitations can be added anatomical specializations. The elephant-trunk-fish or mormyrids have taken their adaptations to specialist diets in turbid water too far for easy modification. Likewise the predatory tiger-fish is stuck with its sharp fangs fixed in large jaws.

Cichlids, on the other hand, have skull and body proportions that are easily altered, and tooth or gut structures that are readily modified. Originally likely to have been litter and plankton feeders that sucked food into the mouth, they process it with a mill-like structure in the pharynx which has an easily modified milling or rasping surface. They have an advanced type of closed gas-filled swim-bladder that requires no priming or gulping of air at the surface. Instead of relying on natural aeration for a superabundance of ova, cichlids invest in smaller numbers of carefully tended eggs. Extrapolating from small riverine cichlids such as *Haplochromis multicolor* (which resembles an Oligocene fossil), the original strategy would have been to guard a clutch of eggs, fan them with the tail and fins and, from time to time, move or clean them with the mouth. From this it is a small step to take the eggs into the mouth every time danger threatens. Many contemporary lake species take the eggs into the mouth immediately after laying, hatch them in the mouth, and only release the young when they can fend for themselves. Some cichlid mouths are so capacious they can accommodate up to 1000 eggs, but most rear much fewer numbers. A very great advantage of mouth brooding is that it emancipates the cichlid from innumerable hazards that limit other types of fish. Mouth brooding, nest guarding, egg turning and cleaning, and fanning with the tail and fins ensure the survival of a very high proportion of young fishes and permits year-round breeding because (unlike some other lake fishes) there is no need to migrate up seasonal streams to spawn.

Territorial aggression and nest guarding are common among fish, but for cichlids protection of a nestful or mouthful of eggs or young is an essential part of their success. Fishermen on Lake Malawi call a particularly fierce species (*Tilapia rendalli*) 'the rogue elephant of the water-weeds' because it rushes out to attack their legs. While sensitivity to intruders varies greatly, most cichlids try to exclude or intimidate other fish at some time in their breeding cycle and this exclusion could play a very significant role in their ability to speciate by throwing up a barrier of aggression around very localized breeding grounds.

In addition to all these advantages, cichlids mature and breed quickly, so there is a very fast turnover of generations. In the race to colonize Africa's lakes, most other fish have been too specialized or too slow to reproduce and so have been left behind. A mere handful of riverine cichlids has exploded into hundreds of lake species. The result has been a series of replays on fish evolution. In each lake the cichlids diverge according to diet: algae and plants, which can be found and grazed in many ways; insects, adult or immature; plankton; crustaceans, dead or alive, on the lake surface, its floor or drifting in between; fish eggs or larvae; larger fish; and molluscs. Each presents problems to be solved, demands the development of special techniques and mouth parts or some odd combination of skills. There are even special 'kiss of death' fish that suck eggs and fry out of their mouth-brooding cousins. Another sucks mayfly larvae out of their burrows with grotesquely swollen lips.

The most obvious manifestation of all these adaptive directions is in the size and position of the mouth and jaws, which enlarge or diminish, swing up or down. A mere increase in size can have a surprising effect on the appearance of a fish. Eyassi and Manyara are two very saline rift-floor lakes which share an endemic cichlid, *Tilapia amphimelas*. In the wet season Eyassi is fed from a much smaller reservoir, the shallow Lake Kitangiri. Here, run-off from the surrounding uplands collects before spilling down the Sibiti River into Eyassi's closed basin.

In Kitangiri, *Tilapia amphimelas* grows twice as long as it does in Lake Manyara, but this is not an overall enlargement. The body's depth and the lower jaw are disproportionately large, giving the fish a very characteristic 'dished' face. In side view they remind me of those breeds of dog or pig with overshot lower jaws.

Another example of changed proportions in a local endemic comes from the middle of Lake Rudolf. It has been 7500 years since the lake was high enough to

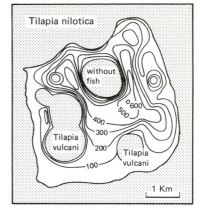

The Island in Lake Turkana where *Tilapia nilotica*, isolated in recently formed volcano craters, have evolved into a distinct species, *Tilapia vulcani*.

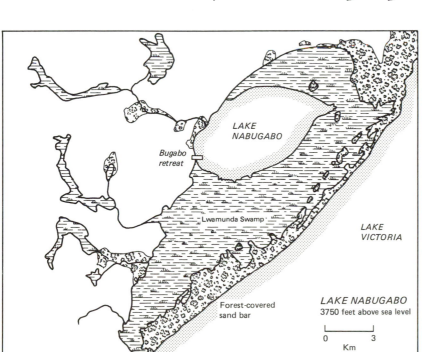

Lake Nabugabo (after a map from the Cambridge University Biological Survey, 1961).

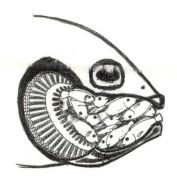

A brooding *Tilapia* cut away to show mouth-brood.

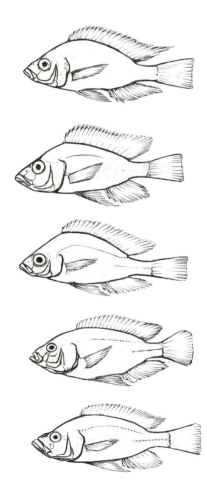

The endemic fish of Lake Nabugabo. *From top,* Haplochromis velifer, H. simpsoni, H. annectidens, H. beadlei, H. venator.

strand a population of *Tilapia nilotica* in one of the volcano's three craters. This is presumably the time it has taken for the Volcano Fish, *Tilapia vulcani*, to become different. At any event, this theoretically 'inbred' population flourishes in what can resemble a bowl of hot pea soup. The crater lake is sometimes so full of algae that it looks completely opaque. Certainly the environment and population size are radically different for the two related Tilapias, although they are separated by only 300 m of lava.

An even more impressive demonstration of evolution behind a narrow barrier comes from a still younger lake. Less than 4000 years ago wind and water threw up a sand bar across the mouth of a bay on the north-western shore of Lake Victoria. The impounded water behind this natural dam wall is known as Lake Nabugabo. Today it is a tranquil sheltered oval of shallow water surrounded by a much more extensive area of swamp. The lake is slowly but inevitably silting up, but the thick masses of vegetation around Lake Nabugabo act as spongy filters so that Nabugabo water is unlike that of Lake Victoria outside. The filtered water of Lake Nabugabo lacks calcium, and without this building material molluscs and crabs are unable to grow their shells and consequently are not found in the lake. The snail-borne disease bilharzia is therefore also absent; it is for this reason that the lake is a popular bathing spot, with a Mission Retreat built on its tranquil northern shore.

Of the nine cichlid fish in the lake, four are indistinguishable from those in the main lake. The other five all have recognizable relatives outside but differ in structure and colouring. For instance, *Haplochromus lividus* in Lake Victoria has a fluorescent blue forehead and brilliant golden flashes on its flanks, whereas its Nabugabo cousin, *H. annectidens*, has scarlet fins and a red lower half. Both species are predominantly grazing herbivores but the choice of foods for the Lake Nabugabo fish is inevitably smaller. It is not only food species that are less diverse in a small lake; there is also less variety in the situations where that food can be found. The teeth of *H. annectidens* show the first signs of its becoming more specialized as a grazer, but its tenancy of Lake Nabugabo is too recent for these modifications to have gone very far.

In each of these instances the colonists have been advanced types of lake fish moving from a bigger into a smaller lake. Although they may have been transformed in their new habitat they remain single sibling species of their progenitor. They are essentially neo-endemics.

The process whereby riverine cichlids become lake fish is the archetypal and more fundamental transformation. This first step cannot be retraced in the Great Lakes because evolution has gone too far. The second and third steps in which they proliferate within the lake are also obscured by the number of species and the complexity of their inter-relationships. Colonization of a lake by riverine cichlids with a subsequent proliferation of species has occurred in what is, on the face of it, a very unpromising locality.

Rising above the bustling town of Kumba, in Cameroon, is a volcano. In its forested crater, which is about 2½ km across, lies a deep lake which is named after the local people, Barombi Mbo. It is not rich in resources. The forest trees around its margins drop leaves, branches, insects and fruit into the water, but the bulk of water in the lake lies still and sterile, without oxygen. Only the first 20 m below the surface is inhabited by fish; yet it sustains 17 species, 11 of which are endemic cichlids.

The lake has been studied by a British Royal Society team in close association with the local fishing community. They showed that the cichlids are well separated in the foods they eat, in the times they are active or breed and in the water zone that they prefer. Six of the cichlids are predominantly vegetarian and are thought to derive from three ancestors. The most specialized of these, *Pungu maclareni*, browses on a sponge that is also endemic to the lakeshore. Another,

North-west end: Incoming water cooler and well oxygenated. Incoming food on or near the surface

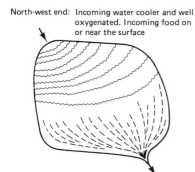

South-east end: Waterlogged and other foods drift towards bottom. Warmer water and less oxygenated towards exit

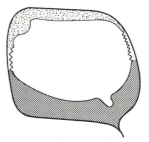

Narrower steeper shores at north end of lake: Broader shallower shores to the south caused riverine fish colonists to diverge into distinct northern and southern populations (separated by narrow rocky zones)

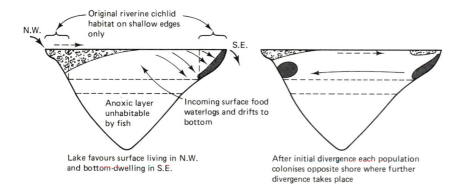

Lake favours surface living in N.W. and bottom-dwelling in S.E.

After initial divergence each population colonises opposite shore where further divergence takes place

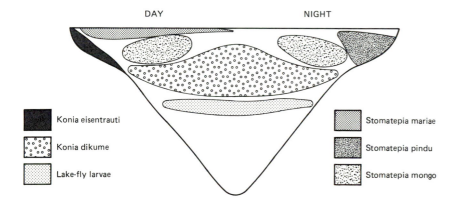

Simplified diagram of zones thought to be preferred by contemporary species

Konia eisentrauti

Konia dikume

Lake-fly larvae

Stomatepia mariae

Stomatepia pindu

Stomatepia mongo

Schematic charts of Lake Barombi.

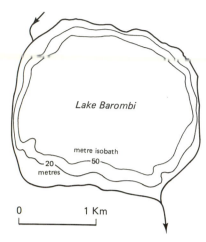

The 20 metre and 50 metres isobaths of Lake Barombi.

Myaka myaka (from the Barombi name, myakamyaka), feeds on phytoplankton in midlake. Five of the cichlids are mainly predatory. They belong to two endemic genera, *Konia* and *Stomatepia*, and they too show separation by feeding zone, choice of food and activity period.

The figure on the previous page illustrates in a very approximate and schematic form how the five predatory species are thought to be distributed in the lake. The six vegetarian species have a more complex distribution in space and time.

There is a particular reason for concentrating on the five predators: they all derive from a single ancestor and they have all speciated within this one lake. It is for this reason that Barombi Mbo is quite the most interesting small lake in Africa. Unlike Nabugabo, it boasts no clear-cut barrier to help explain speciation. In Barombi Mbo the barriers that have made speciation possible were primarily erected by the cichlids themselves, and it is this propensity to erect behavioural barriers between contiguous populations that may explain why the African lakes are populated by such large numbers of cichlid species. Barombi Mbo is therefore the mini-model for a process that is of much wider significance for Africa's inland waters.

Instead of visualizing Barombi Mbo as a homogeneous, symmetrical cone of lake water, consider what a cichlid newly arrived from the river below, would have encountered. A cichlid's earliest venture into a lake would have been severely conditioned by its previous riverine existence; so the fish was likely to have been strictly confined to banks that resembled those of the riverside. The richest pickings in Barombi Mbo are still around its margins, under the over-hanging trees.

This is the point to delineate the real contours of Barombi Mbo (which have scarcely changed since it was first invaded) and to match the lake with the supposed sequence of events that culminated in five fishes evolving from one. Underwater, the crater floor is less symmetrical than it appears from above. The northern half is somewhat steeper than the southern, and it is here that the feeder stream, having drained the forested hills above, empties into the lake. The surface of the northern half of the lake is therefore better aerated and probably richer in nutrients than the southern. Other than vegetation on the crater's rim, this stream has always been the only significant source of incoming nutrients.

The quality of surface water above the wider more gently sloping southern beach may differ only very slightly from that in the north, but differences undoubtedly exist. The fact that free matter is more likely to drift and settle at the lower end of a lake, implies a slight shift in the distribution of nutrients from the surface towards the floor. Subtle as these differences are, they could have been sufficient for the first cichlid colonist to favour more surface-feeding in the north, more bottom-feeding in the south.

Interposed between the two halves of the lake are steep walls where the fore-shore becomes exceptionally narrow and rocky. Such differences in the exact pitch and texture of the crater's walls are insignificant for most of today's 'lake-fish'. For the riverine colonist creeping round the edges, these narrows would have been zones with very few fish. A consequence of this could have been a division into two populations. The southerners would have snaked out on either side of the exit while a separate northern population would have hugged the shores beside the mouth of the feeder stream.

Lake Rudolf and Lake Nabugabo cichlids demonstrate the rapidity with which colour differences can appear and how mouthparts can be modified with different diets. In Barombi Mbo, spatial separation was incomplete but breeding between surface-feeding northerners and bottom-feeding southerners could have been inhibited in a number of ways. The simplest form of dislocation (and one that roughly coincided with the steep rocky sections) would have concerned the daily and annual timing of activity. The best time for feeding and breeding could differ around the mouths of affluent and effluent streams.

A slight difference in the breeding calendar could then be reinforced by colour. Most cichlids are able to modify their colours and there is a general tendency for surface feeders to be lighter in colour, bottom feeders darker. The food preferences of one American cichlid have been found to coincide with particular colour morphs. Such polymorphism could help to reinforce isolation because the cichlid female's choice of a male is influenced by his breeding colours.

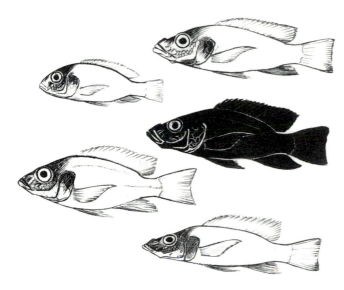

Five Lake Barombi cichlids. *Clockwise from top left,* Konia eisentrauti, Stomatepia mariae, S. pindu, S. mongo and Konia dikume.

How does this hypothetical outline match with the predator cichlids of Barombi Mbo? *Konia eisentrauti* (after the Barombi name Konye) is an inshore bottom feeder, whereas *Stomatepia mariae* is a species that often takes insects from the surface and in deeper water. The first is a dark-coloured fish with transparent fins, the second silvery grey with brassy iridescence and orange fins in the mature male. An especially interesting feature of *Stomatepia mariae* is the marking of young ones with 'St Peter's thumb print', a black midline and tail spot. These markings are very widespread in lake-dwelling Tilapias, especially while they are young and swimming in shoals. They are thought to help in the visual co-ordination and synchronization of movements, and they tend to fade away when the juvenile shoals disperse as the fishes mature.

These two species probably represent greatly modified descendants of the first two predatory fish in the lake. If the north/south division accorded with feeding zones, ancestral *Stomatepia* might have emerged as the northern surface form, *Konias* as the southern bottom feeders. The second phase of their evolution would have occurred as the two distinct species expanded *their* ranges, an event which could have been triggered by external climatic changes. In the encounter that followed, each species would have remained dominant in its region and it should not be supposed that each easily made way for the other. Whereas it had been physical differences between north and south that initiated change in the first colonist, each now had its sibling as a major constraint.

An extrapolation from contemporary niches suggests that the gentler gradients of the south could have meant that *Stomatepia*'s incursion simply squeezed *Konia* inshore. By contrast, the latter's invasion of the steeper northern shores had a different outcome. To avoid competition with the well-established proto-*Stomatepia*, *Konia* went deeper.

If the riverine cichlids' preference for edges was still in operation there would soon have been four distinct populations. Their speciation would have followed similar principles to the earlier divergence except that all four would have continued to change and only when this process was very far advanced would they have become true 'lake fishes' with a continuous distribution all round the lake.

The four emergent species would have been the ancestors of *Konia eisentrauti* (the inshore bottom feeder), *Konia dikume* (deep-water specialist), two shallow-water *Stomatepia*, and *S. mongo* in deeper water. The final divergence of the shallow-water form into a diurnal species (*S. mariae*) and the nocturnal *S. pindu* could have begun with a separation of activity and behaviour between open beach and rock-sheltering populations.

Konia dikume has been the subject of fascinating research which has revealed an exceptional ability to store oxygen by concentrating its very abundant haemoglobin—it is 'the reddest-blooded fish in Africa'. This enables it to dive deep in deoxygenated water to catch lake-fly larvae.

Barombi Mbo provides a tiny microcosm of the evolutionary radiations that have generated 130–190 endemic species in each of the great lakes of central Africa. The shallow scalloped edges of Lake Victoria and its past ups and downs may have encouraged speciation of the Nabugabo type but this cannot be invoked for the deep trench lakes where the Barombi Mbo model seems more relevant. Although the scale is very different, there are actual examples of differentiation in progress at opposite ends of Lakes Malawi and Tanganyika. The occurrence of polymorphism in a Lake Tanganyika cichlid, *Tropheus moorii*, is especially interesting because the 10 or more morphs are scattered along the coastline and show how a linear population can be broken up into sub-populations by a variety of minor interruptions, beaches, rockfalls or sandbanks. The amount of time cichlids have had to diverge has its measure in numbers of endemic genera and in the range of fish sizes (but naturally enough a small lake will tend to have fewer large fish).

As the Rudolf and Kitangiri examples showed, the profile view of a fish's jaws offers us some sort of instant appreciation of adaptive radiation. In the case of Lake Malawi the inter-relationships between genera have been studied and tentative evolutionary trees have been constructed (see fig. opposite). Likoma and Chisamalo Islands in Lake Malawi are foci for dense concentrations of endemic fish. The open water which surrounds the islands separates their rocky shores from similar habitats on the main coastline. This emphasizes how important natural discontinuities between subtly different shorelines are for the speciation of highly sedentary edge-loving fish. The cichlid radiation has been plotted in some detail for Lake Tanganyika, but the relationships are more opaque and the tree even more tentative. Here modifications have proceeded so far that there are tuna-, gurnard- and goby-like fishes, snapper-, barracuda- and grouper-like fishes. Their specialized ways of life are reflected not only in their teeth and jaws but in the entire build of their bodies (see fig. over the page). It is scarcely surprising that the first biologists to see Tanganyika's aquatic fauna thought there must be a sea connection, so faithfully have the forms and varieties of ocean life been recapitulated.

The diversity of cichlids is most immediately apparent in their colouring. The need to escape herons, cormorants, otters, larger fish and innumerable other enemies requires most cichlids to pass a large part of their life in some form of camouflage. Females are often (though not always) drabber than males. Those from open waters tend, like marine fish, to be silver. Shoaling species have prominent lines, spots or bars which probably serve as Stop–Go and orientation signals. Many cichlids, like chameleons, can change their colour very fast, and

● Black
◓ Black/orange intermediates
① Orange 1
⊙ Yellow
◕ Blue-black
② Orange 2
▲ Red and yellow
○ Brown
✪ Green and yellow
✕ T. duboisi

Colour forms of *Tropheus moorei* at the northern end of Lake Tanganyika.

	LAKE			
	Barombi Mbo	Victoria	Malawi	Tanganyika
Age	?	750,000	1,500,000	3,000,000+
Endemic cichlid genera	4	4	20	33
Fish length range (in cm)	3.3–16.5	9–37.5	6–51	3.7–80

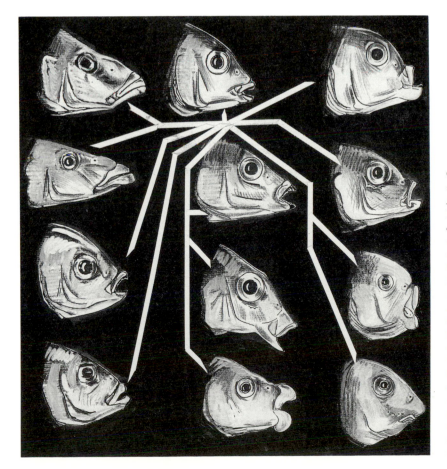

Cichlid radiation in Lake Malawi. From a relatively undifferentiated *Haplochromis* ancestor (probably most like the leaf browser *H. similis, top centre*) a variety of herbivorous and animal-eating types have evolved, such as *Haplochromis compressiceps* (eye-biter) *top left*, and immediately below it *Haplochromus polyadon* (hunter). Zooplankton is eaten by *Cynotilapia afra* (*third row left*) and molluscs by *Haplochromis placodon* (*bottom left*). Larvae caught by different methods are eaten by *Labidochromis brevis, Haplochromis lethrinops* spp. and *Haplochromis euchilus* (*centre column*). Likewise various techniques of plant and algal scraping are practised by *Genyochromis mento, Cyathochromis obliquidens, Petrotilapia tridentiger* and *Labeotro ppheus fuelleborni* (*right hand column*).

Pelmatochromus subocellatus.

this can turn aggression on or off as if at the flick of a switch. Inconspicuous fish burst into a parade of colours, or, in the Transvestite Fish, *Pseudotropheus auratus*, turn from male to female colouring with great rapidity. Others have colours that change more slowly, with age, sexual maturity or by season.

Fins flag a more specific semaphore, and it is on the anal fins that cichlids often carry their most distinctive and significant signal—egg dummies. The males of mouth brooders are faced with a very peculiar problem. The male's presence is necessary before the female will expel her eggs but in some species, notably the Lovelace Cichlid, *Haplochromis burtoni*, she is so quick to gather them up into her mouth that they do not get fertilized. In attempting to gather up male pseudo-eggs, the female gets a mouthful of semen and thus achieves fertilization in the mouth.

Two-dimensional egg dummies on the anal fin are not the only solution to this uniquely cichlid problem. Some species of *Tilapia* dangle a genital tassel that mimics a cluster of eggs, while *Opthalmotilapia* males elongate their pelvic fins and grow two pseudo-eggs (like bait on a fishing line) on trailing fin tips. In some species mimicry of the eggs is very realistic but some anal fin patterns give every sign of being redundant. In the blue and black Zebra Cichlid, *Pseudotopheus zebra*, some of the egg-dummy markings have 'migrated' to bead the fin's outer margin with their yellow spots. Similar spots border the dorsal fin and tail where they can hardly serve the same purpose. In any case, most of them no longer resemble eggs. Egg-dummy patterns demonstrate the cichlids' talent for harnessing colour and pattern to serve specific ends.

The cichlid radiation in Lake Tanganyika. From a very small number of similar river fish, the lake has spawned a variety and range of fishes showing striking similarities with well-known marine types. *Top row, left, Xenotilapia megalogenys, right, Limnochromis leptosoma. Second row, left, Xenotilapia sima, right, Boulengerochromis microlepis. Third row, left, Cyathopharynx lucifer, right, Cyathotilapia frontosa. Fourth row, left, Spathodus marlieri, right, Lobochilotes labiatus. Fifth row, left, Telmatochromis caninus, right, Telmatochromus vittatus. Bottom row, left, Bathybates ferox, right, Eretmodus cyanostictus.*

In the course of invading a new niche or acquiring new habits, cichlids are obliged to modify some detail of a repertoire that is otherwise shared by all species. When such innovations in behaviour can benefit from visual advertisement, the cichlid can readily enlist colour signals in its service. Two sibling populations that encode their behaviour in this way have a potential for jamming each other's communications, thus inhibiting interbreeding. It has been noticed, for instance, that the patterns of closely related lake species frequently differ the most.

The brightest colours often occur in crowded fish communities where they serve courtship or male rivalry. Even so there are exceptions. The female of *Pelmatochromis guentheri* is brighter than the male, and both sexes are an equally bright blue in *Labidochromis caeruleus*. In *Pseudotropheops tropheops* the female is lemon yellow with a black stripe on her dorsal fin while her mate is black with brilliant blue spotting. Any pair of fishes must exchange a sequence of coded messages before eggs can be successfully fertilized and that exchange is literally coloured by the fishes' visual signals.

How exclusive and elaborately staged such signals can be is illustrated by the riverine cichlid *Pelmatochromis subocellatus*. In this species three or more yellow-bordered black spots fill the upper half of the male's enlarged tail fin. Like the ornament on a heraldic banner these very distinctive markings flutter with every switch of the tail. The female has a short functional tail without decoration, yet a single spot with its yellow halo appears on her dorsal fin just for the duration of the breeding season, at which time the male becomes jet black with white flashes. The same species also alters some of its more subtle tints according to mood.

Many of these brightly coloured fish are now popular in aquaria, where their breeding behaviour is being studied.

Among the habitats to which cichlids have adapted are seasonal floodplains, pools and swamps, but these are really marginal and are dominated by those fish that can withstand drought and lack of oxygen, such as lung-fish and cat-fish.

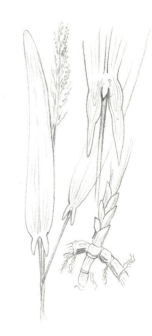

Suddia, a newly discovered giant grass, endemic to the Sudd.

Shoebill, *Balaeniceps rex*

Swamp systems are very extensive in several parts of Africa. The south-eastern headwaters of the Zaïre River are punctuated by numerous swamps and shallow lakes, Bangweulu, Mweru, Upemba and (in Tanzania) Malagarasi. Lake Victoria is surrounded by a network of small valley swamps, especially to the north. The largest swamp in Africa is the Sudd (an Arabic word for blockage) on the Nile, while other major systems are the Okavango, the inland delta of the Niger in Mali and the strings of relict swamps south-east of Lake Chad.

Endemism is not a conspicuous feature of swamps. The Niger–Chad–Nile systems share a very high proportion of their flora and fauna. Marsh birds in particular range very widely, but there are exceptions. The Slaty Egret, *Egretta vinaceigula*, inhabits the Okavango and the upper Zaïre swamps, where it hunts small fish, insects and small snails on grassy floodplains, frequently using its feet to stir up the water. They form small parties of up to eight birds and are active all day. They derive from the same stock as the Black Heron *E. ardesiaca*, which is a very successful and widespread species well known for its habit of catching fish under a canopy formed by its low-spreading wings. Perhaps the Slaty Egret represents a declining cousin of the Black Egret. Like most piscivorous birds in seasonal habitats it must move with the ebb and flow of the floodwaters, and this induces localized movements within its rather restricted range. The much larger Shoebill or Whale-headed Stork, *Balaeniceps rex*, is primarily an equatorial bird from the swamps of the Nile and Zaïre headwaters (but there are isolated records from around Lake Chad).

Shoebills differ from other storks and herons in not fishing their prey along the shallow outer margins of the floodwater. They are permanent residents of the pools, grassy waterlogged meadows and channels deep within large swamps. They avoid papyrus, still stagnant pools and open or deep water. Standing on clumps of vegetation they ambush large fishes, snakes, turtles and the occasional large frog as these aquatic animals travel along the margins of pools or along channels. The water in such places is usually in motion, even if only very slowly. One reason for the Shoebill's restricted range is that such conditions are only found within major swamps on sluggish, but flowing, river systems. Furthermore, it is only in tropical waters that fish are abundant enough to reward such an inactive strategy. The bird is dependent on intercepting travellers on very predictable pathways.

There is another constraint that also helps explain the Shoebill's extraordinary beak. The birds build isolated single nests way out in the centre of the swamps. The nests I have seen were on termite mounds but the birds can also construct their own mounds out of piled weeds. Completely exposed, chicks must withstand great heat and the threat of predators for prolonged periods while the adults wait for prey. The parents need to be reasonably close to the nest so they can look out for predatory birds (which are rare) and return from time to time to dowse the young birds with water. It seems likely that the beak has broadened specifically to bucket up water, and its design achieves this without spoiling its effectiveness as a fishing implement.

In the Sudd the 'meadows' of *Leersia* grass that Shoebills favour are maintained by herds of Nile Lechwe, *Kobus megaceros*. These magnificent antelopes are endemic to the Sudd, where they exist in large herds. Female Lechwe superficially resemble small Red Deer. They are a bright golden colour and possess long pointed hooves. Males are appreciably larger and when fully adult have long sweeping horns and a bold black and white pattern which is the product of progressive darkening over several years. Every harem male expends enormous energy in monitoring and mating females, and in pursuing and fighting any male that comes in sight. Soon exhausted. he is displaced and retires to recuperate alone in the swamps. Seen from an aeroplane, these pied hermits stand out well as they rest within small trampled arenas.

Nile Lechwe, *Kobus megaceros.*

Adult males therefore form a cadre or class whose major concerns, fighting and mating, can be pursued in the midst of a herd of peacefully grazing females. Their pied colouring prevents any confusion in this clear separation of sexual roles. Alone among antelopes, the Nile Lechwe can withstand prolonged periods up to their necks in water, but numerous termite mounds and a few ancient *levées* do provide hauling-out places, even in the depths of the Sudd.

Lechwe and Shoebills tend to have the inner swamps to themselves during the flood season, but as the waters fall many other animals and birds move in. This is also the breeding season and early mornings are noisy with the splashing of young Lechwe as they romp around, running and jumping through the shallows, leaving swirling patterns in mud that has the texture of churned grey oil paint. The Lechwe are soon surrounded by crowds of storks, ibises, egrets and Long-toed Plovers, while hosts of wagtails dance around in pursuit of mosquitoes. Carmine Bee-eaters come in after grasshoppers and the sticky flies brought in by herds of cattle, kob and even the occasional Roan Antelope. At nightfall the Lechwe seek firmer ground or shallows where they are surrounded by the mumbling of crakes and piping frogs and insects. Long-tailed Nightjars chase moths where the pratincoles hunted by day, flying in and out of laminated layers of mist and smoke that drift out from fires on higher ground. Here the Nuer and Dinka herdsmen talk and laugh, enveloped in the smoke from smouldering dung fires. Less traditional fishermen cease all activity at nightfall and retire under white mosquito nets that are pitched on the numerous termite mounds that rise out of the swamp.

It is often not appreciated that the most inhospitable and remote wildernesses in Africa are virtually all inhabited, even if only sparsely, by people. Animals and plants that excite enormous interest among scientists, collectors and naturalists are often a familiar part of such people's day-to-day life and have been for centuries. When field scientists first began to study the fish in Lake Malawi their guides and mentors were the local fishermen. Not only did the fishermen have names for almost every species, they classified the rock-grazers as 'mbuna' and the open water mixed-species shoals as 'utaka', terms which have been adopted by English-speaking naturalists. The fishermen pointed out the predatory habits of *Docimodus*, which takes pieces out of other fish, and the death-shamming *Haplochromis livingstoni* which snaps up smaller fish that have come to feed on the 'carcase'. They discriminated between two very similar fish, Kaduna (which has 16 fin rays) and Kajose (which has 17 and a narrower eye). They pointed out habitats such as the *virundu*, lake-floor dunes where fishes breed; they noted the importance of currents for the fish, 'mkachi' from the south, 'Nkonde' from the north.

Although there is ample precedent for latinizing local names this simple gesture of respect or gratitude for local knowledge and guidance has been uncommon. An exception was the Royal Society/British Museum studies of Lake Barombi Mbo. Here the local expert Joseph Ndokpe Sangwa and the local chief Martin Malleh were acknowledged in scientific publications, and fishes that were 'new' to science kept the names the Barombi had used for centuries, *Pungu*, *Myaka myaka* and *Konia dikume*.

Every year aquaria in Europe and America receive new and exquisite fish varieties from remote parts of Africa, yet their place in the fascinating ecosystems from which they were taken remains largely unknown. The inland waters of Africa remain an unexplored frontier, another unknown in a continent whose darkest secret is human ignorance of its rich natural history.

CHAPTER 12

Man-made Islands—
Reserves and National Parks

Not far above the town of Malakal, on the Nile, there is a long swampy mud-bank called Fanyikyang Island. In the tall reeds and swamp grasses that cover it live several hundred Nile Lechwe. The antelopes spend the day grazing the shallows but gather on muddy rises to spend the night. Even at the height of the dry season most of the island is waterlogged and always lush and green.

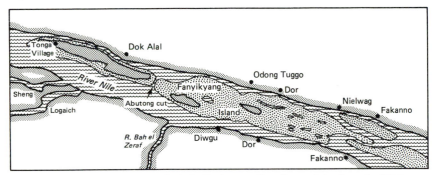

Raised ground and banks of Nile

Mat of inundated vegetation

Rivers, channels and ponds

● (Dor)= Settlements

The traditional hunting reserve for Nile Lechwe on Fanyikyang Island on the White Nile above Malakal town.

By contrast, both banks of the river opposite the island are raised, well-drained and quickly desiccate in the dry season. At this time, the conical grass-thatched huts of the Shilluk villages only house old people and young children with a few sheep and goats. The younger men and women move westwards to find pastures for the cattle along the receding floodwaters of the Nile. Some-times the Lechwe are within earshot of the children as they shout and play under the Borassus palms, and each week the big Nile barges thump and clang past the shores of this long narrow island.

The Shilluk people who inhabit the region have maintained Fanyikyang as a refuge for Lechwe over centuries. It is not a disinterested protection but with-out it there would certainly be no Lechwe there. Every year when the commu-nity is reunited at the end of the dry season, hundreds of men converge on the island in canoes and a long line of spear-carrying beaters wade from one end of the island to the other. Afterwards the canoes are loaded with speared Lechwe and everyone returns to the mainland for nights of drinking and feasting: a sort of Nilotic Christmas. The scent of burning hooves fills the air as naked, fatty Lechwe shanks are grilled on open fires, bones are broken open for marrow and all savour their single annual taste of succulent Lechwe meat.

There are many motives for leaving a tract of land in its wild state. Religious traditions have protected a cycad forest in the Transvaal for two centuries or more. Likewise at Zagné, in the Ivory Coast, villagers protect a grove of relict forest because the mona and spot-nosed monkeys that live there are considered to be holy. Whatever the motives for protecting a tract of wild country, the end effect is to create an island surrounded by humanity.

Africa's natural communities are in the process of being cut up into separate islands and it will become increasingly difficult to distinguish man-made island communities from those that were once naturally isolated. Nevertheless, the distinction will remain a fundamental one and the need to save unique endemic communities from extinction will continue to be a compelling reason for the declaration of new national parks.

While the final chapter will touch upon some of the practicalities of conservation within Centres of Endemism, this chapter considers how parks have come into existence and how people have become exterminators, island makers, and perhaps eventually species makers.

The Shilluk, Venda and Senufo reserves demonstrate that the concept of conserving distinctive pieces of land is ancient and widespread even if the motives differ very greatly. In effect, each of these peoples put a very high value on keeping one small patch of the home territory in a more or less natural state.

International conservation as we know it today has no less local beginnings, but ships, planes and satellites have all helped to enlarge people's perception of what constitutes the 'home patch'. The village conserves its sacred grove, national conservation authorities plan within their own state boundaries, while international bodies concern themselves with 'world heritage sites' and 'global genetic resources'. The vocabulary, perceptions and interests of the three tiers have very little overlap even if they share a common interest in conserving Lechwe, cycads or mona monkeys. The problems this may raise are touched upon in the next chapter. The point to be made here is that although hunting reserves and sacred groves may be as old as mankind, the concept of national parks is very, very new, especially in Africa. It is a concept that is still in an early experimental phase.

The earliest formal, legally designated national parks were essentially the creation of enthusiasts who were also influential lobbyists (such as President Theodore Roosevelt in the USA). One consequence of this was that the siting of early parks tended to be rather arbitrary and rationales for their creation could be quite eccentric, such as saving a view, or a particular game animal. It was no accident that their creators were often military sportsmen.

For example, in 1926 the Sabie Game Reserve, then little more than a politically inspired frontier buffer-zone, became South Africa's first national park as the culmination of an isolated battle campaign waged by Colonel James Stevenson-Hamilton. Likewise the first national parks in Kenya were the outcome of a concentrated crusade, almost an assault, led by Colonel Mervyn Cowie against a reluctant colonial government.

The Serengeti plains.

The Serengeti plains.

Once an arbitrary block of territory has been institutionalized into a national park it quickly takes on in the public's mind and speech an identity not less distinct than a great city or a mountain. As a child I was often taken out in a box-body station-wagon for drives across a game reserve on what people in Musoma called the Ikoma-Itutwa plains. The drier eastern section of these plains lay in Masai territory where they had the more euphonious name Serengeti. An 'Ikoma–Itutwa National Park' might have been more acceptable to Musoma chauvinists in 1951 (when Colonel John Molloy helped persuade the Tanganyika authorities to proclaim a National Park) but such a name would please no one now.

Because the park had proved unpopular with pastoral tribes in the region, the colonial government was on the brink of dismembering it in 1958 when Dr Bernhard Grzimek, the director of Frankfurt Zoo and a celebrated film and TV personality mounted a campaign to save the park. The film he made, 'Serengeti Shall Not Die', was an instant worldwide success. In common speech 'Serengeti' has acquired connotations as vivid as 'Everest' or 'New York'. It is a park and a

One national park viewed from another, Mt Kilimanjaro (Tanzanian N.P.) viewed from Amboseli N.P., Kenya.

name that have probably come to stay. Yet the spectacular migrations of animals that the park's boundaries were meant to enclose can be matched in scale in other parts of Africa while the landscape and setting bear no comparison with eastern Zaïre, Uganda, Kilimanjaro and the southern highlands of Tanzania. The huge success of Grzimek's film and the tourism that followed during the boom years of the 1960s led to the creation of new national parks and reserves in almost every African country. They usually centred on accessible concentrations of large plains or savannah mammals in areas that were not subject to immediate claims for agriculture or settlement. At first little thought was given to making parks truly representative of all habitats but in the 1960s this began to change and the absolute authority of economists and their values began to be challenged.

In early attempts to explore 'supra-economic values' with local audiences (culminating in an inaugural paper to the Uganda Society in 1973) I used to take the central African forest communities to exemplify a search for objective ways of measuring their biological value.

'The simplest means of measuring biological richness is of course compiling lists of species, genera, families and higher taxons and giving them points. However in assessing the richness of a biological community living in a circumscribed area I think an elaboration of the point system is possible. Following the initial assessment of species there should be a further examination of the overall distribution of each species to determine its relative rarity and further points should be given to each species with a restricted range. Still further points could be awarded to species that are exceptionally vulnerable for any reason. Local endemics would, of course, get the highest points.

Measuring wealth is important because feeling wealthy is probably as significant for most people as actually being wealthy and I think that officials, game wardens, foresters and politicians might take more pride in their responsibility to world conservation if they were provided with data that showed them in simple numerical terms what a "biological millionaire" their country is. In addition to such an appeal to pride, status, and responsibility, it should be remembered that there has been in some parts of Africa a tradition of social responsibility in relation to the land where a man saw himself as a mere link in a long

The Murchison Falls in the Murchison Falls National Park Uganda.

chain of life extending way back into the past and on into the unforeseeable future. It must be remembered that the economist can never put a value on a species. All he can do is recognize that a species (or the biological community of which it is a part) is unique and irreplaceable. So what value can we put on a unique form of life?

A species is the realization of a unique possibility of existence. *Our own collective existence as the human race is but one single possibility of existence and each single species is a manifestation of* other *potentialities.*

We ought to be shocked if the authorities in charge of a habitat containing the only living examples of unique species were not conscious of their charge and zealous to execute their responsibility to protect their continued existence. Sadly we must conclude that such ill-informed authorities and administrators do exist and that there is a general ignorance of man's newly found responsibilities. Man now decides on the survival or destruction of other forms of life on this planet and he has a special responsibility towards that unit of value we call the species because we are now in a position to know that many of these species only live or grow in certain strictly limited spots.

It is important to remember the zoogeographic pattern which suggests that some areas are ancient relict refuges while other areas have been more recently colonized by widespread and common dominant species. This pattern reveals that not all species nor all areas of forest are of equal value. Some very small east African forests possess special unique endemics: others possess an incredibly rich variety of rare species and these small areas are often superior in number of species (and conservation points) to those vast forests of Canada, Scandinavia and Russia. The means of preserving representative habitats and giving them a value is through a mosaic of small sanctuaries all over the country under the ultimate control of a governmental Environmental Department but which could often be loosely attached to educational and other agencies.

The determination of the areas which should be conserved would operate at different levels. The first would be the overall national level.

A second level operates within the large game and forest reserves where activities such as pastoralism and logging might be a threat to the survival of particular biological communities or species. Some reserves need to be rated a higher or lower conservation value and the mosaic of small sanctuaries that I believe is essential should be computed accordingly. The size and the frequency of the sanctuaries should take account of the overall value of the area and in important reserves should be determined by certain parameters such as 1) Recognizable vegetation types (ensuring that every type is represented) 2) Catenas (with drainage and altitudinal zones or belts included) 3) Rotations, where cycles can be recognized, there should be a provision for the operation and study of this dynamic element in appropriate strips and protected zones. The average percentage of area given over to mosaic sanctuaries in indigenous vegetation reserves should be about 10 per cent. In areas of peripheral interest or importance this could be diminished, while in very important floral and faunal refuges areas it should be greater.

Looking forward into the not-so-distant future it is not impossible to see legislation obliging large scale developers, be they open-cast miners, oil-men, sugar growers, ranchers, tea-planters or timber mills to devote 10 per cent of their allocated area of exploitation to sanctuaries for the original fauna and flora of the region or an equivalent of their profits to finance less well-endowed areas. The future tithing of large scale consumers and enterprises might be one way of making development directly responsible for these supra-economic values that its activities threaten.'

The ideas we laboured over in the late 1960s in Uganda have now become commonplace and recent years have seen a proliferation of statistical formulas and scoring systems to measure conservation priorities. From a perspective of some 20 years it must be admitted that statistics have proved dull and indigestible fare while a more enduring and infectious enthusiasm surrounds the mystique of the rare. It is a mystique that sometimes resembles the popular reverence for gold.

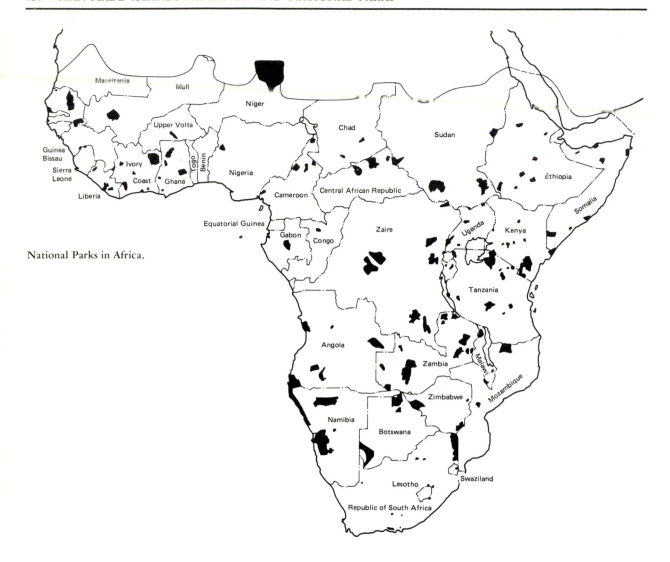

National Parks in Africa.

In 1979 the launch of the World Conservation Strategy signalled the beginning of co-ordinated efforts to put conservation on a rational, systematic and truly global base. The idea of ensuring that every region, ideally every country, should conserve viable areas that are representative of every habitat or ecological zone is therefore a new one. It is a submission that some countries and many individuals reject. That Centres of Endemism within such habitats deserve yet higher priority is an even newer concept, and one that is made less acceptable because such centres are very unevenly, indeed unfairly, distributed. Thus, in a continental perspective poor countries such as Somalia and Tanzania are of much greater importance as wards of Africa's 'genetic resources' than (marginally) richer countries such as Senegal or Zimbabwe, which have relatively unexceptional fauna and flora.

There is now at least one precedent for conserving Centres of Endemism but it comes from a relatively wealthy country with a vast territory. In the last few years Brazil has made Forest Refugia the starting point for identifying and delineating parks within its Protected Areas Plan.

A less systematic but broadly similar approach has already been followed in Zaïre but its adoption will be much more difficult for a poor country such as

Tanzania, which has a fairly dense human population along much of its long coastline. For political and commercial leaders in Tanga only equivalent and immediate benefits could recompense them for the timber and cash-crop revenues that the Eastern Usambaras could produce. (However, it should not be assumed that Tanzania, Africa and the world cannot be woken up to their impending loss in such places.)

Distinguishing between parks that harbour unique endemics and those that sample broader ecosystems may seem invidious. To some extent it is an artificial distinction because all places are unique and the communities that live there are all distinctive to some degree. Nevertheless, as previous chapters have shown, very significant degrees of magnitude are involved. Does the distinction offer any practical insights that could guide the direction of conservation and its priorities? Consider first if there are any parallels between the rather haphazard man-made actions that create islands of wildlife and the natural processes that have generated islands of endemism on particular shores and mountains.

In preceding chapters I have suggested that some of the endemics that are now rare and localized derive from populations that were once widespread and common. The reasons why broader fields have become inhospitable to such species are many: changes in climate, too frequent fires, more competition from other animals (including close relations) and the arrival of new predators and diseases. A habitat may even be closed to an animal because the plants have developed too many poisons.

There is now a single generator of all these forces of exclusion. Until this century people were only capable of making their own islands within the larger habitats of Africa but our generation is witness to the beginnings of a gigantic and completely unprecedented switch-over. Natural habitats are beginning to be the islands in a sea of ranches, plantations, farms and settlements. More than geology, climate or evolution, humans have now become the main creators of biological enclaves. The consequences have never been seriously considered because this biologically cataclysmic event has been the incidental by-product of other concerns. It is the slow, steady result of people with hoes, axes and matches in the fields or the rather faster product of people in bank boardrooms with maps, spectacles and dollars.

It can be argued that our modification of the environment is only one of degree, our biological impact being comparable with a sustained outbreak of locusts or a particularly severe drought. It could also be argued that our impact has only intensified with the passage of time. There is reason to believe that our ancestors reinforced the already destructive effects of extreme drought and cold by killing off the over-specialized large mammals. (This could be predicted because the animals' ranges would have greatly contracted during such periods and these would have been premier hunting grounds for primitive people.) The anatomy of several Pleistocene giants suggests they were unwieldy and probably easy to kill. Their extinction coincided not only with falls in temperature but with the rise of man. At least nine different species of elephant are known to have died out within the period of human history or immediate pre-history; most were actively preyed upon and three of them were African species. At much the same time several giant pigs, a mega-buffalo, three specialized giraffes, three hippos, several equines, sabre-tooths, hyaenas—the list goes on—all have become extinct.

The broader trend in which large mammals become rarer and rarer, continues today. It is exemplified best by the Grass or 'White' Rhinoceros, *Ceratotherium*. In spite of its prehistoric appearance this is not an archaic species but a relatively recently evolved grazing rhinoceros that derived from the same stock as the much older Browse or 'Black' Rhino, *Diceros*.

Grass Rhino skulls and bones turn up with great frequency in a wide scatter of

Grass Rhino, *Ceratotherium simum*

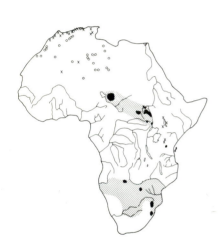

○ Rock art (Genus not always certain)

× Skeletal remains (Genus of record not always certain)

▨ Total recent range (according to historical records, Mauny, 1957; de la Fuente, 1971)

The distribution of Grass Rhino.

late fossil deposits and in archaeological sites. Their portraits are found on rocks in the Sahara, and eastern and southern Africa. They were evidently found throughout the grasslands and savannahs and would have been especially widespread during moister periods (they need to drink daily). Equatorial forests bisected their range into northern and southern populations (which differed slightly). Among rhinoceroses they are unusual in their placidity and willingness to form herds. In 10 BC, a Roman army officer, Julius Maternus, described rhinoceros gatherings in Agysimba (now thought to be the Fezzan country east of Lake Chad). They were still widespread in southern Africa at the time of the first European explorers but by the early years of this century the southern rhinos were estimated to number between 10 and 20. The northern population saw its most drastic contraction more recently, in the 1970s and 1980s, with numbers now down to a few dozen.

Extinction for the southern Grass Rhino was averted by the single-mindedness of one man, B. Vaughan-Kirby, the first Conservator of Game in Zululand, who ensured rigorous protection for the remaining rhinos throughout the 1920s and 1930s. Censuses revealed there were 150 animals by 1929, nearly 1000 by 1960. Their listing as rare and endangered has tended to equate the Grass Rhino with such 'genuine' relicts as Bontebok and White-tailed Gnu. Not so—given a chance they could once again be common and widespread. Their respite from extinction has raised the possibility—in theory if not in practice—that Grass Rhinos could be returned to many parts of their former range.

Few animals could be better suited than *Ceratotherium* to 'island dwelling' (even in quite small areas). Once their relations with humans are secure the animals almost behave like domestic stock. Reintroduction of the Grass Rhino to many reserves has been opposed because of their supposed absence in historic times. Yet there are probably few savannah areas these animals did not once inhabit. It needs to be more widely appreciated that faunas, especially large mammal faunas, were already artificially impoverished long before parks were thought of. We cannot bring back sabre-tooths and chalicotheres but a few of the larger and more ecologically diverse parks could foster a richer spectrum of animals than they do now.

This prospect has become very remote because the rapid extermination of Browse Rhinos throughout the African parks and reserves has demonstrated that few modern parks have the capacity to protect such animals. In fact there is often a discouraging discrepancy between the splendid array of national parks shown in brochures and maps and the plight of large animals within them. Nevertheless, Africa's current parks and reserves are reasonably well distributed throughout the continent and they are broadly representative of the major zones, from desert to rain forest (see p. 238 and the regional maps).

Some of the largest areas contain a broad spectrum of animals and plants with a reasonable chance that they can survive within the park boundaries. In the smaller ones, long-term survival of the larger mammals (and particularly specialist predators and some rare plants) must be in doubt once all the surrounding areas have become settled. For example, few of the woodland parks in western and central Africa will be able to contain genetically viable populations of Derby's Eland, *Taurotragus derbianus*, an antelope that exists at low densities and wanders extensively.

The problems for scarce, widely spaced species finding mates in an artificially constricted area are compounded by the risk of inbreeding. In this respect geneticists and population biologists have come up with a rule of thumb. The minimum population size is 50. At best, this represents a 1 per cent rate of inbreeding which represents the maximum acceptable level without threatening a population's long-term fitness.

On such criteria, a park of several thousand square kilometres would be needed to maintain a population of say, Martial Eagles, each pair of which ranges over 200 sq km. Before the smaller parks are engulfed by settlement, their longer-term prospects will need to be reviewed, especially present and future parks intended for the protection of major Centres of Endemism.

Earlier chapters have established that the ancient refuges have enjoyed a good measure of climatic stability, sometimes for many millions of years. Some species living there have therefore changed very little over the same period and are utterly dependent on a stable predictable environment. Golden moles in southern Africa, otter-shrews in Ruwenzori, guinea- and rock-fowls in west Africa, and the Golden-rumped Elephant-shrew on the Zanj coast are all likely to be sensitive to disturbance or fragmentation of their habitats and populations.

The loss of such species, which belong to unique archaic families with very few living forms, is more serious than the disappearance of regional representatives of common types. Africa has more families of higher vertebrates than Europe and Asia. For example, there are 84 bird families (compared with 67 in the vast Palearctic region and 74 in the Oriental region) and 50 mammal families (compared with a mere 27 in the Palearctic and 43 in the Oriental).

It is known that at least a quarter of the species found as fossils in Pliocene deposits of around 3 million years ago are still with us, although some of them are now rare relicts (such as the Mountain Nyala and Gelada in Ethiopia, or the Okapi in Zaïre). By 2 million years ago the medium–large mammal fauna of Africa was essentially modern. This is in striking contrast with Europe and most of Asia where a modern fauna emerged more recently, with the last of the Ice Ages.

Even the most remote of African parks therefore conserve faunas that are considerably richer and of greater antiquity than those of the northern hemisphere and the Far East. African parks are also important for conserving the habitats in between the moist forest and dry open habitats, these are dynamic ecological zones that have not begun to be adequately explored. There is a particular reason why they should be studied. The fossil record shows that these were preferred habitats for early man and that their basic ecological structure has not changed a great deal since the first emergence of humans in Africa. Many parks contain mosaics of open grasslands, woodlands, forested river valleys and lake shores where southern apes, *Australopithecus* or early forms of *Homo* could probably still make a living were they not extinct. The sense of continuity is reinforced in Serengeti and Lake Turkana by the presence of actual fossil sites within the parks. Here many bones scarcely differ from those of the species still living there. Perhaps even more spectacular, in the Mahali Mountains National Park, in western Tanzania, bands of Chimpanzees range through the savannahs in as close an approximation to early man as it is possible to get without actually resuscitating the dead. Here the apes make very significant seasonal shifts between moist forests, dry open woodlands, thickets and riverine valleys. The steep slopes of the Mahali Mountains are still very largely unexplored.

There are numerous instances of rare plants and animals confined to narrow interfaces (particularly in zones lying between climatic extremes). The minimum area that can support such species is very difficult to compute and choice habitats of this sort are sometimes inadequately represented because they were settled by people before a park was declared and were therefore excluded from the park. The species from such narrow habitats may take a while to die out but their fate in some instances may already be sealed (for example, the Gabela Robin, *Sheppardia gabela*, comes from just such a narrow zone on the Angola escarpment, which is being rapidly settled).

Once the great savannahs or woodlands have been carved up into a few widely separate island parks, some of the animals within them can be expected to change. Alterations in population densities or total numbers can have knock-on effects on sex ratios, social organization and behaviour. Even the anatomy or appearance of a genetically isolated group may change.

As an illustration of the sort of long-term changes that may take place, consider the shape and length of antelope horns. Trophy hunters have long known that animals from particular localities grow larger horns than those from other areas. For example Reedbuck, *Redunca, redunca*, pack in on the eastern *levées* of

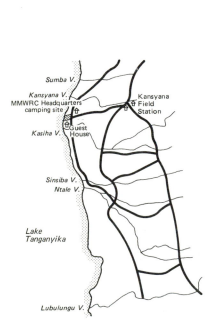

Mahale Mountains National Park.

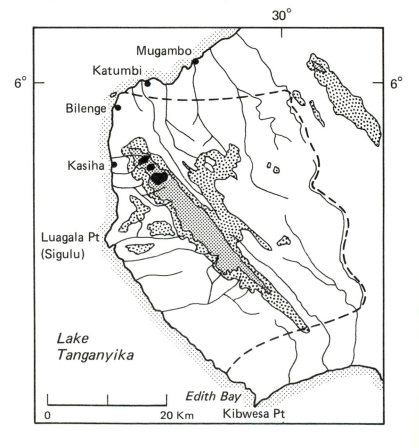

the Nile in south Sudan at up to 63 per square kilometre and the males boast longer horns than those from anywhere else in Africa. In most instances larger horns correlate with higher population densities because more frequent fighting exerts greater selection for stronger weapons. The horns of the Hartebeest, *Alcelaphus buselaphus*, typify those of a high-density antelope. Rutting males spend almost all their time chasing, threatening and fighting (even the females are caught up in the generally high levels of Hartebeest aggression). The heaviest horns growing from an elongated pedicel or skull mounting come from the open savannahs of west Africa where the local race, *A. b. major*, was once one of the commonest antelopes.

In the woodlands of south-eastern Africa, open grasslands occur as small pockets within much more extensive areas of well-shaded woodland. These small glades and valleys support a much smaller and very scattered population of Hartebeest that lives in small groups with little contact. Like the island birds that have dispensed with unwanted muscle and bone, the antelope that carries unnecessarily heavy horns soon selects for smaller ones and this the low-density woodland Hartebeest, *A. b. lichtensteini*, would seem to have done. Cross-sections of their skulls show wrinkled 'collapsing' patterns in the bone that suggest that a once-sturdy pedicel has shrunk down to more modest proportions.

Not only may intermediate populations disappear, leaving long-horned antelopes in one island park and short-horns in another, but many parks will inevitably see fairly consistent alterations in the densities of their animal populations (with all the changes in behaviour and ecology that that entails). Eventually, the animals in different parks could be identifiable from their behaviour and by the shape or size of features such as horns. These alterations will be the direct result of our having fragmented a formerly continuous population. As creators of islands, we will become, willy-nilly, species-makers or at the very least sponsors for new 'park subspecies'. We can expect to see the development of a 'Serengeti race' of gazelles or a 'Golden Gate oribi'.

Earlier on I asked whether it was useful to distinguish between parks in Centres of Endemism and those that sample broader ecological zones. Are there implications for practical policy and are we offered any insights that might guide our priorities in conservation? I think there are. When the terrestrial flora and fauna of Africa are viewed as a whole, the larger part is (within the limits of three major habitats: desert, savannah, forest) adaptable and wide-ranging.

Individuals from low and high density populations of hartebeest have very different horns. *Left*, *Alcelaphus buselaphus lelwel*, *right*, *A. b. lichtenstein*. The inset sketches show the hollow pedestals that underlie the horn bosses. 'Wrinkled' bone in *lichtenstein* suggests a secondary shrinkage or collapse of the pedicel.

During the last million years or so these are the species that have expanded and contracted their ranges with each climatic shift. These could be called 'staple' species to distinguish them from the less mobile species that have remained in stable centres on mountains, coasts or in river basins. The relative importance of endemics *vis-à-vis* generally distributed 'staple' species will vary very greatly from taxon to taxon. Some groups have many isolates, others none, some tend to become regional or ecological specialists, others remain wide-ranging types.

Until this century, three-quarters of Africa's mammal fauna ranged widely within their own life-zones. One-quarter was of very restricted distribution. If these ratios roughly approximate to those of other groups it can be said that about a quarter of the flora and fauna is tied up in enclaves that account for a tiny proportion of the continent's surface. As has been shown in previous chapters, these species are not just relicts. Many are specialists with unique adaptations and biological properties, some are the nearest we are likely to get to 'living fossils' and remembering the many new tools, techniques and concepts that are emerging from microbiology and genetic engineering, the closest we will get to possessing 'time capsules' from Africa's past.

To equate living species with fossils invites the criticism that relict organisms are merely redundant junk and Centres of Endemism are evolutionary scrapyards. The existence of uncompetitive species heading for natural extinction reinforces such a criticism. The same argument is often applied to elephants, rhinos and giraffes. It is a small step from there to claim that all reserves are museums of the obsolete and redundant. I hope that such views will have been offset by previous chapters and by the sketches they contain of an immensely dynamic continent. Here, through combinations of geography, climate and historical accident, substantial traces of Africa's biological history are preserved in enclaves. Numerous and often tiny, these enclaves offer us a vivid display of Africa's diversity, a diversity that many more people would treasure were they to learn of its existence and meaning. In preserving larger and supposedly more robust communities from destruction, human civilization may now be extending and mimicking the natural processes that have already made Africa 'a pattern of islands'.

CHAPTER 13

Island Africa and Its Conservation

Over a short period in 1977–78, Munyarukiko, Sebahutu, Gashagazi and other hunters from the village of Kidengezi on the slopes of Mt Karisimbi, Rwanda, killed and mutilated several Mountain Gorillas that had become trustful of humans because they had been closely observed by researchers over a period of 20 years. The first killing (of an adult male that had been named 'Digit' by the American researcher and writer Dian Fossey) received worldwide publicity. What was not adequately explained was why the Kidengezi villagers killed the gorillas—not for food, not in self-defence and not for profit, for although they axed off heads and hands the taking of these grisly trophies was an afterthought. There is no doubt, however, that the killings expressed the villagers' alienation from official and international efforts to conserve these last Mountain Gorillas. In other Centres of Endemism the outcome could hardly be as melancholy and as immediately tragic, but there are few conservation areas that can wholly escape similar conflicts.

In other parts of Africa, locals resentful of their government's requisition of some part of their territory or usurpation of their rights have speared rhinos, poisoned zebras or put reserves to the torch. Especially at risk are the many localities where an expanding population of poor people are ignorant of the global importance of their own home patch. Expansion curbed by policies shaped elsewhere has always met unequivocal local opposition, whether it was settlers in the Cape in the 1880s, pastoralists in Ngoro Ngoro and Somalia or Virunga villagers.

A dead mountain gorilla.

Acts of vandalism are all too easy in small and precious localities; they signify profound oppositions of interest and values that have no easy resolution. Yet modern societies must address them if the survival of rare species and habitats is to be reconciled with human progress.

Any point where separate interests come into regular contact raises political issues, whether the contact is between individuals or social groups. In controversies over land-use in Centres of Endemism the contestants are generally separated by differences in wealth, class, education, philosophy and culture and by imbalance in power. Any lasting reconciliation is unlikely unless both the issues and the interest groups are properly identified and their positions explored openly and discussed widely.

The centres of endemism that have been described in this book are enmeshed with the lives and subsistence of societies which are impoverished and because they are also expanding need more land and resources. Other claims on land-use are therefore a very real and emotive issue.

Knowledge of rarity and the will to preserve unusual ecological communities depends on education and is largely linked with town dwelling. The commonest ways in which educated urban people can express their concern for nature is by giving their money or their voice to influential conservation societies. The cumulative power of that support is very great, especially since conservation is now woven into the institutions, education systems and even the vocabulary of most countries. Educated classes are therefore likely to support both the philosophy and the policies needed for effective conservation. Interest in conservation continues to grow both within and outside Africa, reinforced by books, films, TV, zoos and gardens. This growth in awareness has spanned continents and cultures but its material existence comprises little more than printers' ink, videotape and a few precarious transplants. By contrast, the living substance of these natural communities resides in the territories and economies of people who can afford few of the media that their rulers, educated classes and the developed world take so much for granted.

The realities of power are exactly the opposite to those perceived by most of the participants in this struggle because the long-term future of Africa's Centres of Endemism lies with local peasantries rather more than with transient governments or enthusiastic conservationists, yet the locals seldom receive the respect that is generally accorded to those that wield power. Meanwhile both populations and resentments grow.

Those who implement conservation in the field have begun to recognize that they derive their immediate power from contact with a wider world and a particular sort of education (and one that is increasingly tempered by biological knowledge). Although they may have effective control over the management of large tracts of land they have no monopoly over the basic questions of who are the ultimate owners of the land, who decides what should be done and to achieve what ends? It is as well to remember that these are questions the poorest peasant asks and for which there are many answers. The conservationists' answers (especially to the last), should not lie in brief propaganda campaigns, which are generally seen for what they are, but in a shared growth of knowledge and debate. The minimal demands of local communities will include sustained not ephemeral programmes of action in which their own people can find meaningful, decisive and dignified roles.

Dialogue between the real parties to the conflict is rare, because the issues, when discussed, and the deals, when struck, seldom involve discussion at the Kidengezi level, in the local villages and schools. Instead they tend to take place between brokers, functionaries of larger national or international organizations. It is also rare for the complexity and multiplicity of local interests to find expression.

This is, so far, a rather pessimistic analysis, yet there are numerous exceptions and in those instances where local people have been invited to share fully in the effort to conserve endemics there has often been a positive response. In the forging of new values and relationships at the local level, conservation has the potential to find fresh purpose and a revived agenda, as an example from Uganda illustrates. It is an example that puts particular faces and names to a particular place, but the people involved typify participants in a larger global struggle.

Mountain Gorillas in Uganda have found their last retreat in a region of deep narrow valleys and high ridges that the farmers outside call Bwindi or Place of Darkness. During the early years of the Protectorate montane forest covered more than 2000 sq km in this region and the British called it the 'Impenetrable Forest', a name that has reinforced the alienation of outsiders from this cool tangled jungle. By contrast the forest was familiar territory to some hundreds of Batwa Pygmies. The Batwa now live along the perimeters of their former estate which has shrunk to 320 sq km. Most of them work for pioneer farmers close to the forest edge, but some still gather honey and other forest produce, or hunt, and a few act as guides and trackers. One man in particular, Bandusya, inherited all his ancestral skills and knows intimately every valley and hill, and the animals and plants found there.

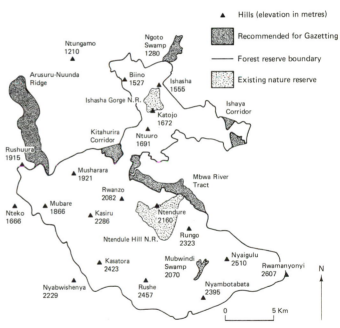

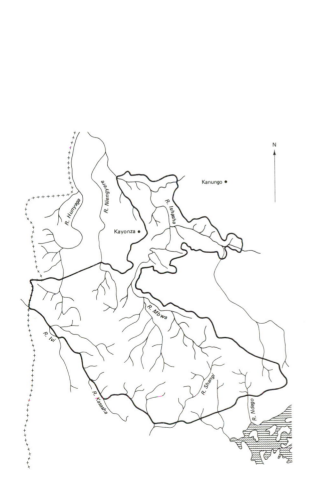

The Impenetrable Forest Reserve, showing the major rivers (left) and the general topographical features (above), including the location of the current nature reserves, the areas recommended for gazetting and the forest reserve boundary.

Bandusya.

For some 40 years visitors to the Impenetrable Forest have depended on Bandusya to show them the forest or its Gorillas. He taught the first game guard, Eriya Bunengo, who is not Batwa and stands nearly two metres tall, not to fear the forest and trained him in the rudiments of bushcraft. Over the years the two of them have acted as guides to senior officials from the capital as well as innumerable naturalists and tourists wanting to see Gorillas and other wildlife. By the mid 1980s after more than a decade of military dictatorship and political chaos the Impenetrable Forest and its 100 or so Gorillas risked obliteration by an advancing tide of peasant farmers, gold miners and timber sawyers. Naturalists in many countries were worried. It looked as though Bandusya's community would witness the final disappearance of the environment that for centuries had given them their living and cultural identity.

When Tom and Jan Butynski first came to work in Uganda in the early 1980s, they became aware of the perilous state and enormous biological importance of the Impenetrable Forest. After several preliminary surveys, in which Bandusya and Bunengo served as guides and informants, the Butynskis wrote a report which persuaded the World Wildlife Fund (and other international conservation bodies) to help the Ugandan authorities with a plan to study and conserve the resources of the region for the benefit of the local people and as a major contribution to local development.

What distinguishes the Impenetrable Forest Conservation Project is a sensitive and ambitious attempt to develop and share new insights and factual knowledge, new skills and techniques, and to promote a network of new contacts between previously insulated interest groups. The project is remarkable for building bridges in a country and in a situation where distrust and mutual ignorance have flourished during two decades of civil war and social chaos. After erecting a field station at Ruhija and instituting all the usual activities of a fledgeling park—such as surveys, research, protection, education and training—the Butynskis and warden Alfred Otim joined local groups to set up a new co-operative project called 'Development through Conservation in South West Uganda' (DTC for short), which aims to bring new tree-planting, soil- and water-conserving skills and a basic environmental awareness to the 86,000 people who live around the margins of the forest. Local and central government bodies, schools, the Red Cross and the National University, have all been enlisted. The research station at Ruhija has become a meeting ground for at least four very different tiers, or classes, with a direct interest in the conservation of the Impenetrable Forest.

Bandusya represents those with the longest and strongest cultural attachment to the area in its natural state. At Ruhija he and his community act as guides, trackers and, informally, as teachers. In many parts of Africa there are peoples with ways of life that rely wholly on the renewable resources of a large territory. Although they have the greatest claims to possession of the land, they tend to be numerically few and are totally unpoliticized and so are powerless in the face of decisions taken at a higher level. Their influence and numbers have been swamped by the expansion of neighbours that are politically organized and demand land for settlement. Such expanding populations normally engulf earlier, smaller communities because the colonists are self-sufficient in crops, livestock and agricultural techniques. So long as they are in their expansionist phase they can always meet their extra needs for such things as timber and recreation, and even water, in the wild lands beyond the frontier or in relict patches. Only when there is no more frontier left do the people have to face up to their need for assured supplies of wood and water. Only when all sources of natural recreation are gone is their loss lamented. Eriya Bunengo, the pioneer farmer, hunter and former game-guard, is representative of the actively colonizing expansionist sector of local society. Bandusya occasionally reminds his former colleague Bunengo that their grandfathers were mortal enemies and that the Batwa once kept all strangers out of the forest. The contemporary reality is that Bandusya's people are subjugated and the Impenetrable Forest is the last vestige of wild land; the last frontier. The knowledge and past confidence of small hunting and pastoral communities makes a sorry contrast with their present degradation. A class of people who might be expected to have had a natural inheritance in today's parks have undergone a total social and political eclipse, whether they are Khoisan in South Africa, Mursi in Ethiopia or Ndorobo in Tanzania. The Impenetrable Forest Project is one of the few enterprises where there could be some vestige of continuity. Here Bandusya's family may possibly find new roles.

The dominant ethnic group in any locality normally exerts some political influence in central government but their concerns are not necessarily identical. The main agents of change tend to be innovative entrepreneurs in agriculture and industry, timber and mining, who are constantly seeking out new areas for expansion. Because they are off the beaten track, wild peripheral lands become anarchic cockpits for competing interests. Kigezi is no exception. The 'Development through Conservation' project has tried to combat this anarchy by inviting local authorities to participate in a forum that seeks to reconcile local, national and international interests, and to anticipate the long-term prospects. Trained to take the longer view and with a knowledge of natural processes, Ugandans with an ecological education are now taking up the responsibilities of leadership. Didas Turinawe, Alfred Otim, Samson Werikhe, Alex Muhweezi and Stanley Bazaakabona all participate in the environmental education programme of the DTC. In the capital, influential leaders such as university professor Fred Kayanja and biologist Eric Edroma have kept environmental and conservation issues in the forefront of public and government concern, lobbying, organizing seminars and serving on crucial policy-making committees. If Bandusya and Bunengo personify truly parochial and involuntary participation in the conservation scene Kayanja, Edroma and others represent new cadres of nationally minded conservationists. Tom and Jan Butynski personify the influence of a still wider world. When this assortment of very different people sit down with one another, or with other local or national authorities or citizens, they bring together an extraordinary mixture of skills and insights. The 10-year 'Development through Conservation' programme provides a forum for the resolution of conflict, where each of the major social and political levels can be represented.

Turning from the particular to the general, there is, at first glance, no shortage of organizations devoted to conservation and the environment in Africa. Most are very new and represent the response to a sudden reversal whereby humans, once islanders in a sea of wild things, are now the sea around islands of wilderness.

Throughout their existence in Africa, people have had to protect themselves from nature. In less than a generation the roles have been reversed, so it is scarcely surprising that the ideas of conservation come into frequent conflict with the still dominant preoccupation of taming wild lands with new techniques and more people. The idea that conservation should make still further efforts and give priority to 'Centres of Endemism' has been resisted by local populations as unwarranted interference in their freedom to use their own local resources. The most frequent source of contention is less that the disputed resources are essential to survival than that a park is an external imposition, yields few benefits to local people and represents one more surrender of local autonomy. All too often these are real and justified complaints, and exhortations from governments and large international conservation organizations will not be heeded unless the needs of local people are met and their enthusiasms enlisted.

In many parts of Africa, modern ideas have superseded prudent traditions that have stood the test of trial-and-error over good times and bad. Yet the locals increasingly find themselves unable to reach the sources of new knowledge (through limited literacy, a shortage of books, schools or a variety of social obstacles) and to participate fully in modern institutions. So the various manifestations of the New Order, good, bad or indifferent, tend to become discredited. Unable to understand the rationale or values behind policies that effect them, and resentful of their exclusion from decisions, they either lock up their creative energies and become apathetic, or they become destructive. In their relations with the environment it is a destructiveness that is often compounded by loss of land and traditions.

Those who would conserve Centres of Endemism will need to combine ecological knowledge with a respect for the views of rural residents, a responsiveness to their needs and a concern that they get a fair deal in future development. This is sometimes made more difficult by people in higher tiers of the local hierarchy. Anything that might diminish their already small powers is resisted, so the poor can be further insulated from any benefits on offer. A major problem facing conservation in these areas is therefore a big gap in the perceptions, education, motives and values of most rural people and those of them for whom the reasons in favour of conservation seem obvious. The latter are still a very small minority. Biogeographers, the first to identify Centres of Endemism, were seldom locally born and bred, so the concept itself begins as an alien one. It is also an alien idea in countries that lack true endemics.

Chief Martin Malleh of Barombi, with Dr Ethel Trewavas of the British Museum and Joseph Sangwa.

The absence of public concern for the survival of African Centres of Endemism is easy enough to explain. For a start, most people say: 'centres of what?' Neither the word nor the existence of such centres is familiar. As for the public valuing these communities, why should they? Values are cultural responses and they emerge from the sort of education people grow up with. So, how are young people, especially residents in these areas, to acquire the knowledge of ecological and evolutionary principles that would give that information a meaningful context? For that matter, how can their governments, agencies and planners become more familiar with the practical help ecology offers in maintaining sustained production and conserving precious resources?

The Impenetrable Forest Conservation Project is one such experiment, but it must be remembered that such initiatives are subsidiary to a network of well-established authorities, which have more formal institutional roles in the conservation of Centres of Endemism, these include the day-to-day administration, maintenance and protection agencies without which all reserves would disappear. For a start, there are those national bodies or large-scale landowners that control or affect the practice of conservation. National Parks, Wildlife, Forestry, Natural Resources and Water Departments may variously have direct responsibilities but Ministries of Agriculture, Tourism, Defence and Power often play a decisive part in perennial debates over land-use priorities and policy, as do the universities and various associations of landowners. Most countries have unofficial societies or pressure groups variously called Wildlife Clubs, Natural History Societies, Ecological Societies, Nature Foundations or Field Sports Groups; (a list of some these societies is given in Appendix 4). Most of them welcome overseas membership and in a country where the government is not concerned about conservation, overseas interest becomes one measure of how seriously the country's flora and fauna is taken.

At the international level, the United Nations Environmental Programme (UNEP) includes among its functions a Global Environmental Monitoring Service (GEMS), which aims, among other objectives, to inventory and monitor centres of genetic diversity or endemism. The United Nations Educational, Scientific and Cultural Organisation (UNESCO) has also been active in conservation and ecological studies (for example, in 1978 UNESCO published the monumental 'Tropical Forest Ecosystems. A state of knowledge report'.) The International Union for the Conservation of Nature (IUCN) has several programmes that are specifically geared for endangered biota and habitats. Among its many initiatives are the Species Monitoring Unit, Threatened Plants Unit and Plants Conservation Programme. IUCN has also published the Red Data Book and the Centres of Plant Diversity Conservation Strategy, which is part of the World Conservation Strategy.

IUCN is closely linked with the most influential of all non-governmental organizations (NGOs), the World Wide Fund for Nature (previously the World Wildlife Fund). At the present time WWF has projects in several countries containing Centres of Endemism and it spends about two-thirds of its entire budget in Africa. In Uganda, Rwanda and Zaïre education programmes on the environment, Gorilla conservation and the expansion of conservation initiatives (such as the proposed Okapi National Park in the Ituri Forest in Zaïre) are under way. Sierra Leone, Liberia and Ivory Coast are being assisted in the operation of new national parks. In Cameroon and Gabon training in primate and rain forest conservation is being supported and systems of protected areas are being reviewed. Tanzania's moist forests, the Uzungwas and Kilimanjaro are the subjects of conservation programmes. Wildlife conservation staff are being trained in Somalia, and the proposed Bale National Park in Ethiopia also receives assistance. In Ghana and Tanzania youth leadership in forest conservation is also supported by WWF. The WWF has an older sib, the Fauna and Flora Preservation Society, FFPS, which also serves as funder, 'linkman' and 'honest-broker' over many conservation issues and was the founder of the Mountain Gorilla Project.

National societies that have played an important part in African conservation are the New York Zoological Society (NYZS), the Frankfurt Zoological Society (FZS), the African Wildlife Foundation (AWF), Conservation International and the Centre for Conservation Biology (CCB). The CCB is attached to Stanford University, which is one of the foreign universities to maintain productive working relationships with African universities. The Centre also represents part of a new concern to bring systematic study to bear on the management of endangered ecosystems. In association with Conservation International, CCB is now helping to train local specialists in rain forest resources management.

One of the problems inherent in all charities, which raise their funds by appeals and advertisements, is that they tend to shape their activities and make decisions with fund-raising 'success stories' in mind. There is a risk that the simplistic outline of a media story told for propaganda gets to be believed by its authors. Whenever publicists claim an organization has saved a species or a locality from destruction it needs to be remembered that it is the local forces *capable* of that destruction that are as significant a power as the savers. Restraint

Uganda botanist John Kasanene and zoologist Isabirye Basuta discussing their field programmes with Lysa and Tom Struhsaker, all members of the Kibale Forest project.

is always a local achievement which should be acknowledged. Credibility and good standing soon get lost in resentment when a conservation body 'hogs' the credit for a project, however laudable their real role. Conservation bodies are learning to moderate their claims. They no longer trumpet 'we have saved the . . .', but instead tend to say 'we have helped so-and-so to make life safer for . . .'. Another alienating trait that professional conservationists have not yet learnt to avoid is uncritical use of the vocabulary of Power. By framing their programmes on terms set to them by economists and political planners they often distance themselves from would-be sympathisers and clothe their concepts in gobbledegook.

Among the many examples of outstanding co-operative enterprises mentioned in this volume the Kibale Forest Project should be mentioned. A tropical research programme was set up in 1969 at a field station in Kibale Forest with funds from the New York Zoological Society and with the support of Uganda's Forest Department, Makerere and Rockefeller Universities. Nearly 20 years of systematic data have been accumulated and many student courses as well as doctoral programmes in ecology and primatology have been pursued there. Scientific papers, books and films have continued to emerge from one small corner of a country that has been torn by civil wars, poverty and the brutality of General Amin's regime. Such achievements could only be accomplished by an alliance of very determined people. Behind the scenes are outstanding Ugandans such as Professor Fred Kayanja, ecologist Eric Edroma, Forest Conservator Peter Karani and two American scientists, Tom Struhsaker and Lysa Leland.

A new generation of Ugandans has now taken over; Isabirye Basuta and John Kasanene both have doctorates that were researched and written at Kibale, and the work of these and other scientists have made this forest a by-word in tropical rain forest ecology.

There are other instances of conservation going on during periods of intense social upheaval. For example, in Ethiopia a number of influential individuals and a disciplined Wildlife Department have succeeded in maintaining numerous fine reserves and national parks. Dense populations can also co-exist with national parks when the people are well enough informed on the benefits and interest of their wildlife. For example, much has happened in Rwanda since the death of Digit and the other Gorillas. The last killing was in 1983 and the proportion of young apes has risen to 48 per cent of the population. An opinion survey has suggested a real change in local perceptions because the local authorities (with support from the Mountain Gorilla Project, the International Centre for Conservation Education and numerous other bodies) have taken a basic environmental education to virtually every school in the country. Among many other lessons, people are learning the simple statistic that a single year's increase in human population would normally require a new area of farmland four times that of the Parc des Volcans (where all the Gorillas live) and that the Park's value as water-catchment area, conservation area, national monument and money-earner (from tourism) far outstrips its immediate worth as agricultural land. (After repeated excisions the Park represents a mere 0.5 of Rwanda's area but no further cuts are likely now.) A broad-based environmental education is backed up by publications and books in local languages such as Nicole Monfort's *Inyamaswa zonsa zo mu Rwanda* an illustrated guide to the ecology of Rwanda's mammals (published by the Kigali Rotary Club).

Even in rich metropolitan countries it is hard enough to find accurate and interesting books on the ecology of African communities. For information on unique local flora and fauna (especially in a local language) *Inyamaswa* is very much a pioneering venture.

Local awareness of and involvement in the uniqueness of an area depends on a basic foundation of knowledge and information. There is a great need for databanks for each of the major Centres of Endemism. At Victoria in Cameroon a very fine arboretum and herbarium, originally founded by German botanists at the turn of the century, had become rather run down before Cameroon set up an integrated conservation and renovation programme in association with the Royal Botanic Gardens at Kew. The scientific study of forest resources and plant conservation will be combined with research, education and administration. Nairobi and Cape Town, by contrast, have world famous and active herbaria. Many more are in the course of being refurbished or developed in other parts of Africa. Overseas, the major sources are the herbarium at Kew and the Plant Sciences Data Centre set up in Washington by the American Horticultural Society. The Missouri Botanical Gardens has developed special programmes of botanical exploration and collection in Africa. Museum collections exist in most capitals, the largest being in Nairobi and Pretoria. Overseas, the American, British, Tervuren (Belgium) and the Paris Museums of Natural History are major sources of preserved material.

Outside interest and support for local conservation begins with knowledge of a locality's global significance. Here lies the rationale and the motive. It should therefore be axiomatic for successful conservation in such areas that the local people who live close to the national parks and reserves should be well informed on the uniqueness of their area. It also seems obvious that wherever possible the teachers, researchers, guides and wardens concerned with that area should be recruited from the local population. Furthermore, the larger part of the revenue from regional parks should go back to the local people. Sadly it is rare for any of these common-sense provisions to be met, and the problem is partly due to a peculiar colonial legacy. National parks grew out of Game Reserves and the Game Departments grew out of elephant control, which was part of the revenue collection for early colonial coffers. Ever since then many central governments and their treasuries have refused to relinquish the principle that monies earned from wildlife, whether from ivory sales, licences, tourist fees and so on, should go to the central exchequer. Redistribution of the income has always reflected the current priorities of the government in power, defence, industrialization, livestock development, etc. Parks and reserves have therefore tended to be starved of cash. Here is one more instance of local people being deprived of the incentive to value what they have. The conservation of wildlife inside or outside Centres of Endemism is as much a political and social issue as it is an environmental one.

At the present time detailed knowledge of the numbers of endemics, of the finite extent of Centres of Endemism and an awareness of their biological or global significance is severely restricted, even where that knowledge exists. If possession of that knowledge has helped generate the impulse to conserve shouldn't such knowledge be shared with those who live within or close to these centres? In an equitable social and educational system the locals, of all people, should be the most knowledgeable and most interested in their own wildlife and in the uniqueness of their own locality. Yet such information seldom comes their way and the frameworks of biological and ecological knowledge that would make such facts and figures interesting and relevant are even more distant.

A large part of the rarity value that these obscure places and species have acquired has come from esoteric and foreign subcultures such as trophy hunting. Wholesale transfer of such foreign values is neither likely nor generally desired. So where is the common ground? What there is a need for and a place for in virtually all cultures is an ecological education based upon local biological realities. The natural context for learning about unique endemic communities is within such programmes.

As with any living body of knowledge such programmes must be sustained by continuous practice, the accumulation of relevant facts and information and by the growth of theory—on how local organisms, including people, interact with each other and with the environment and climate. Those who live where the action is are best placed to enjoy and contribute to that growth. Their fellowship with others will progress from the need to develop conceptual and scientific frameworks that deliver results as well as from intellectual, aesthetic and, it is hoped, economic rewards. The established conservation bodies could do much more to help share and disseminate knowledge, more to put people in touch with one another, and more to increase awareness of the uniqueness of people, places and species.

Conservation is very much an issue of education. Authorities lack information on the resources they are responsible for, and they lack knowledge of the high value of those resources. A large part of the problem is that ecology is too new a discipline to have been part of the education of those who take decisions. As a result, policies and actions, at all levels, continue to operate within narrow terms of reference. The foundation of prestigious international bodies such as UNEP and IUCN has begun to change this and they, together with public opinion, have even begun to influence the policies of major financial institutions. The World Bank, USAID, the EEC and Asian Bank now require environmental and social impact assessments for the projects they support, but it was their custom until very recently to fund enormous development projects with scant regard for their effect on resident societies, local economies and special environments.

The World Bank is the most powerful institution in Africa, loaning many billions of dollars a year and employing 2700 staff, of which 3 are trained ecologists. Favouring giant projects, the bank is technocratic, secretive and a law unto itself, allowing minimal contributions from or information to those local people that are said to be the ultimate beneficiaries. In the Awash Valley of Ethiopia uncounted wild animals and 150,000 Afar pastoralists lost their grazing lands and a further 20,000 people were displaced into camps; the people regarded this as punishment by God. In reality there is no divine control—the board of directors is dominated by the votes of nominees from five Western nations and there is no accountability to recipient communities or tax-payers in the Western world. As with any other bank, the World Bank runs at a profit, much of it deriving from interest and loan repayments from African countries. Its policies have only begun to respond since the clamour of protests became impossible to ignore. The impact of World Bank projects on gorillas in the Virungas, on flora and fauna in the southern highlands of Tanzania, and on wildlife in the Kalahari were ignored initially with disastrous results. It is to be expected that the World's Bank's reintroduction of large-scale plantations to São Tomé and Principe will adversely affect the indigenous flora and fauna. Typically there is no provision for study of the impact of pesticides, clearing and agricultural development on the rich endemic communities of those islands. In spite of improvements, the World Bank still tends to stay within narrow definitions of 'development' and has remained primitive in its scope, preferring crude quick-return 'conversion' into livestock, silviculture and cash crops to less destructive forms of sustainable production. None of the largest financial institutions of the contemporary world has attempted to make a large-scale systematic assessment of the biological productivity of indigenous African ecosystems. Nor have they made any attempt to understand range ecology in any context but that of domestic stock. The development of education, recreation, ancillary services and enterprises in or around the African Parks has also had to look to other sources.

Tropical enterprises geared to high production and profits suffer frequent setbacks due to unexpected attacks from diseases, parasites and predators. To date the automatic response has been 'fire-brigade' counter measures. It can be predicted that more and more of these problems will arise, yet few of the giant enterprises, such as livestock or tropical agriculture, have invested in knowledge of indigenous ecosystems nor have they co-operated in the preservation of locally adapted communities, despite their obvious value as repositories of species.

In 1987, four of the largest financial and development organizations in the world put together a plan to co-ordinate the development of tropical forest resources on a long-term sustainable basis. The Food and Agriculture Organization (FAO), the United Nations Development Programme (UNDP), the World Bank (IBRD) and the World Resources Institute (WRI) have combined to initiate a Tropical Forestry Action Plan which 'offers an opportunity to end the present state of alarm' about the destruction of rain forests. There is still a very wide conceptual gap between 'Economy' and 'Ecology' which is evident in the language the plan uses to classify forests; 'fallow', 'productive', 'unproductive', 'unmanaged', 'logged over', are very different categories to those of the ecologist. Were the Tropical Forestry Action Plan to live up to its promises it could represent a turning point in the relationship between world financial institutions and the environment. The outcome could be significant for the conservation of forest refuges or Centres of Endemism in tropical Africa, because money could eventually be working *for* conservation, rather than *against* as has generally been the case.

Before accepting the Tropical Forestry Action Plan at face-value those concerned with conservation should require assurances from its sponsors. Projects should be formulated openly with free access to information. The ecological aspects of all projects should carry more weight and require that many more ecologists are assigned for independent study of the implications; self-sustainable projects should have priority and the local people should have real control over projects that affect them to ensure that they are the prime and ultimate beneficiaries.

The forest Centres of Endemism have been casualties of the general tendency to lump forests into one category, and there is still too little recognition that forests vary greatly in their biotic diversity. Ecologically there is no measure of the knock-on effects for other organisms when a forest is felled or pulped. What is known is that each time there is a major perturbation and the absolute quantity of accumulated nutrients is reduced the balance will tip further against the less competitive or more specialized species. Most especially the rarer, old endemics will suffer.

Among the many effects of felling and silvicultural management is not only a reduction in the number of species but a radical alteration in the age-spread of trees. Old or dying trees are almost invariably the richest in special niches, their hollows conceal flying mice, bats, owls, lizards, spiders and insects, their rotting branches are loaded with epiphytes and a host of agents of decomposition are at work on their dying limbs (and they can take half a century over dying).

The wasteful exploitation of tropical forests and pitifully low prices paid for tropical hardwoods has led to calls for the nations with tropical hardwood to form a cartel similar to the oil producers'. In spite of its potential benefits, individual rain forest nations continue to undervalue this resource and sell it off cheap. Large timber multinationals are permitted to exploit tropical rain forests with minimal regulation and without regard for whether their fellings are in ecologically sensitive areas, or within Centres of Endemism. Sustainable logging is both possible and desirable over substantial areas, but the activity should be much more carefully and systematically controlled as well as zoned to exclude critical areas within the Centres of Endemism. In the meantime, those with the well-being of forests in mind may do well to question timber multinationals as to what proportion of their profits they reinvest in conservation and field research. Without substantial contributions towards the maintenance of the habitat from which they benefit (say 10 per cent of their profits) they are nothing but asset strippers. For example the timber firm SILETI is currently felling the last important forest area in Gola, Sierra Leone. Current major multinationals involved in the tropical hardwood trade are: Georgia Pacific, Unilever, Mitsui, Toyomenka, Diamond Corporation, Honshu Paper, Ogi and Parsons and Whitmore (who have eight sawmills in Africa and only two in the USA).

It is increasingly important to identify the profiteers from large-scale exploitation of the environment. The individuals or corporations who benefit from such enterprises may be part of larger trends, but it is no longer enough to be fed statistics about how many football-pitch-sized patches of forest fall every minute or to be told that mankind is engineering its own doom. If the places and species that are being destroyed can be identified, so too can the precise interests and people that are destroying them, whether these are local politicians, leading businesses or foreign multinationals.

For all the international involvement and central government authority, the imposing titles and the confusing initials, earlier chapters will have made it evident that there is still a role for individuals, especially for those able to share their enthusiasm with others. In fact, it is a major contention of this book that it is the cumulative impact of innumerable individual initiatives and partnerships at the most local level that will have the greatest impact. Otherwise the sheer scale of the task is sufficient to quickly exhaust the resources of all the great organizations, corporations and institutions put together.

There is much scope for sympathetic help and expertise from abroad. The development of National Parks in many countries of the world has, in a very short time, generated administrative structures, transport and communication systems, ancillary industries, shops and clubs: in short, a microcosm of society itself. With it come new problems and new rewards, and a proliferating need for people with all sorts of special skills, from computer operators and pilots to trackers and nurses. Mostly these needs will be met by citizens, but there is always scope for sharing, comparing, updating and above all mutual learning.

Visitors, especially knowledgeable ones, such as keen horticulturalists, birdwatchers and amateur naturalists, can be at one and the same time learners and mentors, as long as they spurn tours that are insulated from all local contacts. It should be possible for such visitors to link up with local clubs and schools where they can socialize and attend or even give talks. Sponsorship of rare species or even individual animals (such as is practised in many Western Natural History Societies) could help people overseas to feel they had an intimate connection with a far away place and its people while helping out in a country chronically short of revenue. If you can adopt a duck that migrates between Siberia and an English estuary, why not follow the fortunes of 'Hodari', the clever one, a Gorilla on the slopes of Mt Karisimbi, or 'purchase' a hectare of forest in Korup, Cameroon. 'Twinnings' and exchanges between schools, universities, clubs and societies could help Africans to visit each other and overseas to see for themselves the many social and ecological dimensions of conservation. Meanwhile, counterparts from temperate countries may come to appreciate the richness of tropical ecosystems compared with their own lands so recently reclaimed from glaciers.

The large conservation societies could give more encouragement or provide the sort of 'clearing-house' facilities that would help small local societies maintain themselves and keep in touch with a wider conservation world. There is ample room and nothing to stop Wildlife and Natural History Societies making exchanges and forming a network of affiliations right across Africa and beyond. There would be many experiences to share, lessons to be learnt, new friends to be made and fresh fields to discover. The large number of energetic members enrolled in Natural History Societies often allows quantities of data to be collected on, say, bird or primate censuses or seasonal flowerings. Regular surveys and ecological studies can link up keen amateurs with museums, universities and professional conservation biologists to their mutual benefit. The small cadre of professionals that currently attempt to promote conservation could do much more to encourage and co-ordinate multiple small initiatives by amateur groups.

One day Africa may be able to realize its enormous human, natural and mineral resources in a more ordered and humane world in which its natural riches are better appreciated. Such days may seem far away in the upheavals of the present, but the effort to conserve Africa's wonderful array of natural communities can bring people closer. All those involved will develop insights into human as well as biological relationships. Their rewards will not only lie in contributing to survival but in fellowship between different peoples with different skills, each contributing to a richer, more generous relationship with nature. Above all will be the experience of places and wild things of matchless beauty and interest.

Typical Endemic Species in the Islands and Continental Enclaves of Africa

OFFSHORE ISLANDS

INDIAN OCEAN ISLANDS

Seychelles and Aldabra

Plants 233 indigenous species—72 are endemic
One endemic family—Medusagynaceae, *Medusagyne oppositifolia.*
Endemic genera include: *Vateriopsis Geopanax, Protarum* and six palms—
Deckenia, Lodoicea, Nephrosperma, Phoenicophorium, Roscheria, Verschaffeltia
Invertebrates
  Butterflies 301 species—233 are endemic to the Seychelles (81
  genera—12 are endemic) including *Gideona* (1 sp.), *Heteropsis* (2 spp.),
  Sinerina (1 sp.), *Saribia* (3 spp.), *Tichiolous* (2 spp.), *Perrotia* (6 spp.) and
  Arnetta (3 spp.)
  Spiders 46 endemic species
Amphibians Endemic family of frogs (Sooglossidae) with 2 genera and 3
species. Endemic caecilians (*Grandisonia, Hypogeophis, Praslinia*)
Reptiles
  Tortoise Dipsochelus gigantea (Aldabra Giant Tortoise)
  Snakes Boaedon geometricus
  Lycognathophis seychellensis
  Ramphotyphlops braminus
  Lizards Janetaescincus, Pamelaescincus and
  Phyllodactylus inexpectatus
Birds Seychelles Kestrel *Falco araea*, Seychelles Scops Owl *Otus insularis*,
Seychelles Warbler *Acrocephalus sechellarum*, Seychelles Magpie-robin
Capsychus sechellarum, Seychelles Black Paradise Flycatcher *Tersiphone
corvina*, Seychelles White-eye *Zosterops modestus*, Toc-toc *Foudia sechellarum*,
Flightless White-throated Rail *Dryolimnas cuvieri aldabranus*, Aldabra
Warbler *Nesillas aldabranus*, Aldabra Drongo *Dicrurus aldabranus*.

Socotra and Abd el Kuri

Plants Specific to Socotra: *Anglalanthus* (1 sp.), *Bullochia* (3 spp.),
Dendrosicyos (1 sp.), *Haya* (1 sp.), *Lachnocapsa* (1 sp.), *Lochia* (1 sp.),
Mitolopis (1 sp.), *Nirarathamnos* (1 sp.), *Placopoda* (1 sp.), *Socotranthus* (1
sp.), *Trichocalyx* (2 spp.)
Birds Scops Owl *Otus socotranus*, Socotra Bunting *Fringillaria socotrana*,
Sunbird *Nectarinia balfouri*, Rock Starling *Unychognathus frater*, Proto-
cisticola *Incana incana*.

Zanzibar and Pemba

Birds Pemba Green Pigeon *Treron pembaensis*, Pemba White-eye *Zosterops
vaughani*, Pemba Scops Owl *Otus pembaensis*.
Mammals Pemba Fruit Bat *Pteropus voeltzkowi*, *Cephalophus adersi*
(Zanzibar near-endemic).

Comoros

Birds Grand Comoro Scops Owl *Otus pauliani*, Grand Comoro Drongo
Dicrurus fuscipennis, Mayotte Drongo *Dicrurus waldeni*.

ATLANTIC OCEAN ISLANDS

Tristan da Cunha

Birds Inaccessible Rail *Atlantisia rogersi*, Tristan Bunting *Nesospiza
dacunhae*, Grosbeak Bunting *Nesospiza wilkinsi*.

Gulf of Guinea Islands (Annabon, São Tomé, Principe, Bioko, Fernando Po)

Plants *Anisophyllea cabole, Drypetes glabra, Mesogyne henriquesii, Polyscius
quintasii*

*Discoclaoxylon occidentale, Tabernaemontana stenosiphon, Trichilia grandifolia
Peddiea thomensis, Podocarpus mannii* (mostly São Tomé)
Many epiphytes probably endemic
Birds Dwarf Olive Ibis *Bostrichia bocagei*, Maroon Pigeon *Columba
thomensis*, São Tomé Scops Owl *Otus hartlaubi*, São Tomé Fiscal Shrike
Lanius newtoni, Fernando Po White-eye *Speirops brunneas*, São Tomé
White-eye *Zosterops ficedulinus*, Moller's warbler *Prinia molleri*, Newton's
Shrike *Lanius newtoni*, São Tomé Weaver Finch *Ploceus grandis*, Brown
Finch *Poliospiza rufobrunnea*.

Cape Verde

Plants *Euphorbia tuckeyana, Echium hypertropicum, Phoenix atlantica*
Birds Large Billed Lark *Alauda razae*, Reed Warbler *Acrocephalus
brevipennis*.

St Helena

Plants More than 1100 species and many ferns belonging to 28 genera—
38 are endemic
8 genera confined to St Helena: *Commidendrum, Pladaroxylon, Nesiota,
Trimeris, Mellissia, Petrobium, Melanodendron, Lachanodes*
Dicksonia arborescens St Helena Tree Fern
Invertebrates
  Coleoptera Approximately 256 species—137 are endemic
  Labidura herculeana St Helena Giant Earwig
Birds St Helena Plover *Charadrius sanctaehelenae*

Ascension

Plants *Euphorbia origanoides, Sporobolus durus, Hedyotis adscensionis*

CONTINENTAL AFRICA

THE CAPE

Plants 7000–8000 species—more than 3500 are endemic
7 endemic families:
Bruniaceae—12 genera, 75 species
[Including *Brunia* (7 spp.), *Raspalia* (16 spp.) and *Audonia* (1 sp.)]
Geissolomataceae—1 species (*Geissoloma marginata*)
Grubbiaceae—2 genera, 5 species
[Including *Grubbia* (4 spp.)]
Penaeaceae—5 genera, 25 species
[Including *Stylapterus* (8 spp.) and *Penaea* (3 spp.)]
Retziaceae—1 species (*Retzia capensis*)
Roridulaceae—1 genus, 2 species
[Including *Caridula* (2 spp.)]
Stilbaceae—5 genera, 12 species
[Including *Stilbe* (6 spp.)]
Characteristic endemic genera and species:
Ericaceae—18 genera, ca. 600 species
Proteaceae—13 genera, ca. 320 species
[Including *Leucosperum* (47 spp.), *Protea* (100 spp.), *Leucadendron* (80 spp.),
Faurea (5 spp.)]
Rutaceae–Diosmeae—10 genera, ca. 150 species
[Including *Agathrosma* (135 spp.)]
Lithops—37 species
Notable endemics include: *Diastella buckii*, *Erica* spp., *Agapanthus* spp.,
Amaryllis, Dimorphotheca spp., *Ixia, Gladiolus, Kniphofia, Pelargonium
acerifolium, Moraea loubseri, Orothamnus zeyheri, Restio aeockii, Monbretia
Crocosmia* hybrid, *Arctotis stoechadifolia, Asparagus plumosus*
(28 species are extinct, 96 endangered, 125 vulnerable, 336 critically rare
and 500 a cause for concern according to IUCN)

Invertebrates Includes
Phylum: Onychophora—*Peripatopsis* (12 spp.)
Hepialid moth *Leto venus*
  Butterflies Endemic species include: *Thester, Durbania, Dira* and *Neita, Phasis* spp. and *Poecilmitis thysbe.*
  Dragonflies Metacnemis augusta, Enallagma polychromaticum
Amphibians 8 *Breviceps* species, *Leptopelis xenodactylis, Arthroleptella lightfooti, Arthroleptis troglodytes, Anhydrophryne rattrayi, Cacosternum capense, Xenopus gilli, Hyperolius horstocki, Heleophryne rosei.*
Reptiles
  10 species of tortoise including:
  Grooved Mountain Tortoise *Homopus areolatus*
  Geometric Tortoise *Psammobates geometricus*
  Snakes 4 genera, *Dasypeltis, Leptotyphlops, Lycodonomorphus,* and *Lamprophis*
  Chameleons 10–12 species of *Bradypodion* including:
  Bradypodion setaroi (Coast dune forest, Zululand)
  Bradypodion ventrale (Eastern Cape)
Birds Bothas Lark *Spizocorys fringillaris,* Bush Blackcap *Lioptilus nigricapillus,* Cape Rockjumper *Chaetops frenatus,* Orange-breasted Rockjumper *Chaetops aurantius,* Knysna Warbler *Bradypterus sylvaticus,* Victorin's Warbler *Bradypterus victorini,* Cape sugarbirds *Promerops* (2 spp.), Orange-breasted Sunbird *Nectarinia violacea,* Drakensberg Siskin *Serinus symonsi,* Cape Siskin *Serinus totta,* Rudd's Lark *Mirafra ruddi,* Southern Bald Ibis *Geronticus calvus,* Black Harrier *Circus maurus,* Blue Crane *Anthropoides paradisea.*
Mammals Duthie's Golden Mole *Amblysomus duthiae,* Zulu Golden Mole *Amblysomus iris,* Van Zyl's Golden Mole *Cryptochloris zyli,* Giant Golden Mole *Chrysospalax trevelyani,* Cape Golden Mole *Chrysochloris asiatica,* Dwarf Shrew *Suncus varilla,* Cape Elephant Shrew *Elephantulus edwardi,* Cape Dune Mole-rat *Bathyergus suillus,* Cape Mole-rat *Georychus capensis,* Saunders' Vlei Rat *Otomys saundersiae,* Cape Spiny Mouse *Acomys subspinosus,* Bushman Rabbit *Bunolagus monticularis,* Bontebok *Damaliscus dorcas,* Grysbok *Raphicerus melanotis,* Grey Rhebok *Pelea capreolus,* Bluebuck (extinct) *Hippotragus leucophaeus.*

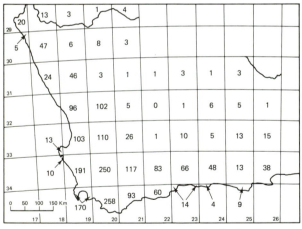

Numbers of rare and threatened plant species in the Cape.

NAMIB, NAMAQUALAND AND KAROO
(South West Africa (Namibia) and The Republic of South Africa)

Plants Approximately 1700 endemic species
1 endemic family—Welwitschiaceae (1 sp.—*Welwitschia bainesii*)
Characteristic families are:
Asclepiadaceae: Stapelieae—ca. 160 endemic species
Aizoaceae (Mesembryanthemaceae—ca. 150 endemic species
Notable endemic genera include: *Adenolobus, Ceraria, Grielum, Phymaspermum, Lithops* (40 spp.). Notable species include: *Aloe pillansii, A. buhrii, A. erinacea, Pachypodium namaquanum, Encephalartos caffer, Latifrons woodii, Stangeria eriopus, Charadrophila capensis* (Cape gloxinia).
Invertebrates
  Butterflies Endemic species include: *Colotis lais, Acraea* spp., *Epamera obscura, Spindasis modesta, Castalius griqua, Sarangesa querdesi.*

  Dragonflies Orthetrum rubens.
Amphibians *Cacosternum namaquense, Breviceps* (2 spp.), *Temoptera tuberculosa, Rana* (2 spp.), *Ptydadena* (4 spp.), *Hildebrantia ornata.*
Reptiles
  Snakes Endemic species include: *Psammophis, Bitis peringueyi, Telescopus* spp., *Leptotyphlops.*
  Lizards Many endemic species including: Sundevall's Lizard *Prosymna sundevallii,* Ocellated Sand Lizard *Uma notata,* Sand Shoveller *Aporosaura anchietae,* Webfooted Gecko *Palmatogecko rangei,* barking gechos *Ptenopus* spp., Brain's Blindworm *Typhlosaurus braini,* Legless Sandskink *Fitzsimonsia brevipes.*
  Chameleons West Coast Chameleon *Bradypodion occidentale,* Damara Chameleon *Bradypodion damaranum,* Karroo Chameleon *Bradypodion karroicum.*
Birds Burchell's Sandgrouse *Pterocles burchelli,* Namaqua Sandgrouse *Pterocles namaqua,* Karoo Lark *Mirafra albescens,* Dune Lark *Mirafra erythrochlamys,* Karoo Bustard *Eupodotis vigorsii,* Blue Bustard *Eupodotis caerulescens,* Ruppell's Bustard *Eupodotis rueppellii,* Black Bustard *Eupodotis afra,* Gray's Lark *Ammomanes grayi,* Herero Chat *Namibornis herero,* Buff-streaked Chat *Oenanthe bifasciata,* Cape Pipit *Macronyx capensis.*
Mammals De Winton's Golden Mole *Cryptochloris wintoni,* Short-eared Elephant Shrew *Macroscelides proboscideus,* Namaqua Dune Rat *Bathygerus janetta,* Mountain Ground Squirrel *Xerus princeps,* Dassie Rat *Petromys typicus,* Pigmy rock mice *Petromyscus* spp., *Gerbillus* (3 spp.), Large-eared Mouse *Malacothrix typica,* Brown Hyaena *Hyaena brunnea,* Mountain Zebra *Equus zebra,* Yellow Mongoose *Cynictis penicillata.*

SOMALI CENTRE OF ENDEMISM (Almost wholly within the Peoples Republic of Somalia)

Plants (Note: the Somali Centre of Endemism is often considered to extend south as far as Tanzania. Under this enlarged definition there are approximately 2500 species of which about half are endemic)
1 endemic family—Dirachmaceae (1 sp.—*Dirachma socotrana;* also in Socotra)
Notable endemic genera include:
Chionthrix, Erythrochlamys, Harmisia, Hildebrandtia, Kelleronia, Sericocomopsis, Xylocalyx, Livistona spp. (Fanpalm)
Numerous endemic species of:
Acacia, Boswellia, Commiphora, Crotalaria, Maerua, Indigofera, Ipomoea, Stapelia
Notable endemic species include: *Euphorbia cameroni, Whitesloanea crassa, Cordeauxia edulis* (Yeheb Nut), *Ceratonia oreothamna, Lavendula somaliensis*
Invertebrates
  Butterflies Endemic genera include: *Dopydodigma, Iolaus, Spindasis, Lepidochrysops, Bicyclus* and *Acraea* spp.
  Beetles Include: *Arthrodebius* spp. of tenebrioid beetles.
Birds Djibouti Francolin *Francolinus ochropectus,* Somali Little Bustard *Eupodotis humilis,* Somali Pigeon *Columba oliviae,* Somali Long-Clawed Lark *Heteromirafra archeri,* Warsangli Linnet *Acanthis johannis,* Grosbeak Canary *Serinus donaldsoni,* Obbia Lark *Calandrella obbiensis,* Somali Bush-lark *Mirafra somalica,* Phillips Crombec *Crombec phillipsi.*
Mammals Simonetta's Golden Mole *Chlorotalpa tytonis,* Somali Hedgehog *Erinaceus sclateri,* Beira *Dorcatragus melanotus,* Dibatag *Ammodorcas clarkei,* Silver Dik-dik *Madoqua piacentini,* Speke's Gazelle *Gazella spekei,* Pelzelns Gazelle *Gazella pelzelni,* Speke's Gundi *Pectinator spekei,* Hirola (a marginal endemic) *Beatragus hunteri.*

THE UPPER GUINEA (mainly within Liberia, Ivory Coast and Ghana)

Plants One endemic family Dioncophyllaceae – *Triphyophyllum peltatum*
Many plants are western forms of main forest types such as *Didelotia unifoliata, Tieghemella africana, Tetraberlinia tubmaniana, Cynometra leonensis, Maschalocephalus* spp., *Diospyros* (7 spp.).
Invertebrates
  Butterflies Species include: *Papilio menestheus, Salamis cytora.*
  Dragonflies Agriagrion leoninum, Allorhizucha campioni.
Amphibians
  Frogs Many species including: *Kassinia lamottei, Nectophrynoides occidentalis, N. liberiensis Hyperolius wermouthi, Phrynobatrachus liberiensis, Phrynobatrachus ghanensis.*
Birds Gola Malimbe *Malimbus ballmanii,* White-breasted Guineafowl *Agelestes meleagrides,* Wattled Cuckoo-shrike *Campephaga lobata,* Senegal Flycatcher *Batis senegalensis,* Nimba Flycatcher *Malaenornis annamarulae,* White-necked Picathartes *Picathartes gymnocephalus,* Yellow-casqued

Hornbill (near endemic) *Ceratogymna elata*, Rufous Fishing Owl *Scotopelia ussheri*.
Mammals Dwarf Otter-shrew *Micropotamogale lamottei*, Liberian Mongoose *Liberiictis kuhni*, Pygmy Hippopotamus *Choeropsis liberiensis*, Gambian Mongoose *Mungos gambianus*, Pardine Genet *Genetta pardina*, Johnston's Genet *Genetta johnstoni*, Royal Antelope *Neotragus pygmaeus*, Jentink's Duiker *Cephalopus jentinki*, Banded Duiker or Zebra Antelope *Cephalophus zebra*, Green Colobus *Procolobus verus*, Diana Monkey *Cercopithecus diana*, Lesser Spotnosed Guenon *Cercopithecus petaurista*, Western Mona Monkey *Cercopithecus campbelli*, Splendid-tailed Squirrel *Epixerus ebii*, Slender-tailed squirrel *Allosciurus aubinni*, Defua Rat *Dephomys defua*, Pel's Anomalure *Anomalurus peli*.

BIGHT OF BIAFRA (Lowland forests Cameroon and S.E. Nigeria)

Plants Approximately 8000 species
1 endemic family: Medusondraceae (Lepidobotryaceae, Octoknemaceae, Pandaceae, Pentadiplandraceae, Scytopetalaceae are other more widely distributed forest belt endemic families).
Notable genera with many endemic species: *Erismadelphus*, *Aframomum*, *Costus*, *Cola*, *Diospyros*.
Invertebrates Very large numbers of insects, notably butterfly *Charaxes acreoides*.
Amphibians Many species including 2 caecilians; 2 *Xenopus* frogs; ca. 20 ranids including Goliath Frog *Conraua goliath* and several species of *Cardioglossa*; ca. 20 species of treefrogs (rhacophorids).
Reptiles Turtles *Kinixys evosa*, *Kinixys homeana*, *Pelusios niger*.
5 species of *Hemidactylus* geckos and *Lygodactylus conrani*, *Diplodactylus weileri*.
Several endemic species of snakes including: *Dipsadohoa isotepis*, *Chameleo widerscheimi*.
8 skink species and several geckos.
Birds Dja River Warbler *Bradypterus graueri*, Yellow-footed Honeyguide *Melignomon eisentrauti*, Bates Weaver *Ploceus batesi*, Grey-necked Picathartes *Picarthartes oreas*.
Mammals Cameroon Wooly Bat *Kerivoula muscilla*, Red-bellied Monkey (marginal) *Cercopithecus erythrogaster*, Sclater's Guenon *Cercopithecus sclateri*, Satanic Colobus *Colobus satanus*, Mandrill *Mandrillus sphinx*, Drill *Mandrillus leucophaeus*, Flightless Scaly-tail *Zenkerella insignis*, Pygmy Squirrel *Myosciurus pumilio*.

ZANJ—EAST AFRICAN COAST AND ARC MOUNTAINS (Eastern Tanzania, Eastern Kenya, Malawi and Mozambique)

Plants Approximately 3000 species—several hundred are endemic.
Endemic genera: *Cephalosphaera*, *Englerodendron*, *Grandidiera*, *Stuhlmannia*. 92 endemic tree species in eastern Kenya and northern Tanzania including: *Greenwayodendron suaveolens*, *Uvariodendron pycnophyllum*, *Cynometra* (3 spp.). Notable species include: *Englerodendron usambarense*, *Allanblackia stuhlmannii*, *Ficus usambarensis*, *Aningeria pseudoracemosa*, *Cola scheffleri*, *Aframomum laxiflorum*.
Among endemics on Kitulo Plateau, Mt Rungwe and neighbouring uplands in the Southern Highlands are: *Tephrosia* spp. (including *T. lepida*, *T. iringae*), *Impatiens* spp. (including *I. flammea*, *I. leedalii*), *I. austrotanzanica* and *I. cribbi*), *Swertia curtoides*, *Ceropegia* spp., *Cynanchum rungwense*, *Streptocarpus* spp., *Vernonia* spp., *Moraea* spp. (including *M. callista* and *M. tanzanica*), *Kniphofia kirkii*, *Disa* spp. (including *D. ukingensis*), *Habenaria* spp. (including *H. goetzeana*), *Halothrix* spp., *Satyrium comptum*, *Diaphananthe* spp., *Polystachya geotzeana*, *Stolzia leedali*, *Kotschya* spp.
Invertebrates East African mountainous region endemic invertebrate genera include: Diplopoda (35 spp.), Sphecidae (27 spp.), Mollusca (55 spp.). (Terrestrial)
Dragonflies *Amanipodagrion gilliesi*.
Butterflies Endemic genera from the coastal belt include: *Graphium*, *Appias*, *Euxanthe*, *Charaxes*, *Neptis*, *Hypolimnus*, *Acrea*, *Pentila*, *Teriomina*, *Epamera*, *Belenois*, *Mulothris*, *Bicyclus*, *Coeliades*, *Cymothoe*, *Uranothauma*, Diurnal Moth *Chrysiridia croesus* and Hawk Moth *Rufoclanis mccleeryi*.
Amphibians Currently known endemics from Tanzanian Mountain ranges—15 spp.
Leptopelis uluguruensis, *L. parkeri*, *L. barbouri*, *L. vermicularis*, *Afrixalus uluguruensis*, *A. sylvaticus*, *Hyperolius spinigularis*, *Phlyctimantis keithae*, *Probreviceps* (4 spp.), *Callulina* spp. *Spelaeophryne* spp., 4 spp. of *Nectaphrynoides*.

Reptiles
Snakes Endemic species include: *Vipera hindii* (Kenya highlands), *Atheris ceratophorus* (Usambara), *Atheris barbouri*, *Atheris desaixi* (Mt Kenya).
Chameleons *Chameleo anchietae*, *C. deremensis*, *C. fuelleborni*, *C. incornutus*, *C. laterispinis*, *C. spinosus*, *C. tempeli*, *C. tenuis*, *C. werneri*, *C. schubotzi*, *Rampholeon brachyurus*, *R. brevicaudatus*, *R. platyceps*, *R. temporalis*.
Birds Usambara Weaver *Ploceus nicolli*, Bertran's Weaver *Ploceus bertrandi*, Clarke's Weaver *Ploceus golandi*, Spot Throat *Modulatrix stictigula*, Blackcap Bush-shrike *Malaconotus alius*, Sharpe's Longclaw *Macronyx sharpei*, Rufous-winged Sunbird *Nectarinia rufipennis*, Taita Thrush *Turdus helleri*, Fulleborne's Robin *Alethe fulleborni*, Hinde's Pied Babbler *Turdoides hindei*, Swynnerton's Robin *Swynnertonia swynnertoni*, Usambara Ground Robin *Dryocichloides montanus*, Anomalous Ground Robin *Dryocichloides anomala*, Iringa Ground Robin *Dryocichloides lowei*, Sharpe's Robin *Sheppardia sharpei*, East Coast Robin *Sheppardia gunningi*, Sokoke Pipit *Anthus sokokensis*, White-chested Tinkerbird *Pogoniulus makawi*, Usambara Eagle Owl *Bubo vosseleri*, Sokoke Scops Owl *Otus ireneae*, Uluguru Bush-shrike *Malaconotus alius*.
Mammals Uluguru Shrew *Myosorex geata*, Kilimanjaro Shrew *Myosorex zinki*, Kenya Blind Shrew *Myosorex (Surdisorex) norae*, Yellow-rumped Elephant Shrew (incipient species) *Rhynchocyon cirnei chrysopygus*, Red and Black Elephant Shrew (incipient species) *Rhynchocyon cimei petersi*, Abbott's Duiker *Cephalophus spadix*, Ader's Duiker *Cephalophus adersi*, Lesser Pouched Rat *Beamys hindei*, Eastern Tree Hyrax *Dendrohyrax validus*, Usambara Squirrel *Paraxerus vexillarius*.

ETHIOPIA (Peoples Revolutionary Republic of Ethiopia)

Plants No endemic families
Endemic genera include: *Butyrospermum*, *Haematostaphis*, *Pseudocedrela*.
Notable species include: *Rosa abyssinica*, *Adhatoda schimperana*, *Impatiens rothii*, *Helichrysum citrispinum*, *Plectranthus edulis*, *Erythrina burana*, *Lythrum hyssopifolium*, *Primula verticillata*, *Aloe* spp.
Invertebrates
Butterflies *Papilio aethiops*, *Bicyclus aethiops*, *Charaxes phodus*, *Lepidocrysops guichardi* (endemic to Bale mountains), *Acraea* 3 spp.
Amphibians *Nectophrynoides osgoodi* and *N. malcolmi*
Ptychadena cooperi and *P. erlangeri*
Leptopelis gramineus and *L. ragazzii*
2 spp. of *Tornierella*
Reptiles
Snakes *Pseudoboodon lemniscatus*.
Birds Sidamo Long-Clawed Lark *Heteromirafra sidamoensis*, Ankober Serin *Serinus ankoberensis*, Abyssinian Cat-bird *Parophasma galinieri*, Prince Ruspoli's Turaco *Tauraco ruspoli*, Black-winged Lovebird *Agapornis taranta*, Yellow-fronted Parrot *Poicephalus flavifrons*, Wattled Ibis *Bostrychia carunculata*, White-collared Pigeon *Columba albitorques*, Thick-billed Raven *Corvus crassirostris*, Blue-winged Goose *Cyanochen cyanopterus*, Golden-backed Woodpecker *Dendropicos abyssinicus*, White-tailed Swallow *Hirundo megaensis*, Banded Barbet *Lybius undatus*, Abyssinian Long-claw *Macronyx flavicollis*, Ruppell's Chat *Myrmecocichla melaena*, White-winged Cliff-chat *Thamnolea semirufa*, White-billed Starling *Onychognathis albirostris*, Black-headed Forest Oriole *Oriolus monacha*, White-backed Black Tit *Parus leuconotus*, Rouget's Rail *Rallus rougetii*, Yellow-throated Seed-eater *Serinus fluvigula*, Black-headed Siskin *Serinus nigriceps*, Spot-breasted Plover *Vanellus melanocephalus*, Stresemann's Bush Crow *Zavattariornis stresemanni*.
Mammals Ethiopian Wolf *Canis simensis*, Walia Ibex *Capra walii*, Mountain Nyala *Tragelaphus buxtoni*, Gelada *Theropithecus gelada*, Dinsho Shrew *Crocidura baileyi*, Grass Rat *Arvicanthis blicki*, Lovats mouse *Dendromus lovati*, Harsh-furred Rat *Lophuromys melanonyx*, Giant Root-rat *Tachyoryctes macrocephalus*.

EQUATORIAL HIGHLANDS

Central African Archipelago
(Ruwenzori, Virunga, Migula, Mitumba Mountains)
Plants [There are approximately 4000 species (approx. ¾ endemic) in Afromontane archipelagos as a whole]
Typical endemic families in Afromontane archipelagos include: Barbeyaceae, Oliniaceae
Genera with endemic species include: *Balthasaria*, *Ficalhoa*, *Kiggelaria*, *Leucosidea*, *Platypterocarpus*, *Trichocladus*, *Xymalos*, *Afrocrania*, *Ardisiandra*, *Cincinnobotrys* and *Stapfiella*
Notable endemic species include: *Chrysophyllum pruniforme*, *Grewia*

mildbraedii, Maesobotrya purseglovei, Xylopia staudtii, Cola bracteata, Tabernaemontana odoratissima.

Invertebrates

Butterflies Species from Kivu-Ruwenzori include: *Papilio leucotaenia, Graphium gudensi, Mylothris ruanda, Gnaphodes grogani, Bicyclus* spp., *Charaxes opinatus, C. furnierae, Euryphura vansomereni, Acraea* spp. and numerous Lycaenidae and some Hesperidae.

Amphibians Species include: Itombwe Frog *Callixalus pictus,* Copper Frog *Chrysobatrachus cupreoniteus.*

Reptiles

Chameleons Chameleo adolfifriderici, C. schubotzi, C. xenorhinus.

Birds Dusky Crimsonwing *Cryptospiza jacksoni,* Shelly's Crimsonwing *Cryptospiza shelleyi,* African Bay Owl *Phodilus prigoginei,* Grauer's Green Broadbill *Pseudocalyptomena graueri,* Lake Lufira Weaver *Ploceus ruweti,* Golden-naped Weaver *Ploceus aureonucha,* Yellow-legged Weaver *Ploceus flavipes,* Black-lored Waxbill *Estrilda nigriloris,* Green-headed Sunbird *Nectarinia alinae,* Regal Sunbird *Nectarinia regius,* Marugu Sunbird *Nectarinia prigoginei,* Copper Sunbird *Nectarinia purpuriventris,* Kabobo Apalis *Apalis kabobensis,* Kibale Ground-thrush *Turdus kibalensis,* Schouteden's Swift *Apus schoutedeni,* Ruwenzori Turaco *Tauraco johnstoni,* Grauer's Cuckoo-shrike *Coracina graueri.*

Mammals Ruwenzori Otter-shrew *Micropotamogale ruwenzori,* Ruwenzori Horseshoe-bat *Rhinolophus ruwenzori,* Hill's Horseshoe-bat *Rhinolophus hilli.*

Angola

Birds Angola Francolin *Francolinus sweirstrai,* Angola Turaco *Tauraco erythrolophus,* Gabela Helmet-shrike *Pricnops gabela,* Montieros Bush-shrike *Malconotus monteiri,* Angolan Thrush *Sheppardia gabela,* Pulitzer's Longbill *Macrosphenus pulitzeri,* White-headed Robin-chat *Cossypha heinrichi,* Loango Weaver *Ploceus subpersonatus.*

Cameroon Mountains

Amphibians

Montane amphibians Bufo preussi, Wolterstortia parvipalmata, Hyperolius kohleri.

Reptiles

Montane lizard Makuya maculilabris
Chameleons Chameleo quadricornis, C. montium, C. pfebleri.

Birds Mount Cameroon Francolin *Francolinus camerunensis,* Double-spurred Francolin *Francolinus bicalcaratus,* Bannerman's Turaco *Tauraco bannermani,* Mount Kupe Bush-shrike *Malaconotus kupeensis,* Mountain Sunbird *Nectarinia ursulae,* Mountain Babbler *Lioptilus gilberti.*

Mammals Preuss's Mountain Monkey *Cercopithecus preussi,* Cameroon Mountain Squirrel *Pavoxerus cooperi.*

ZAÏRE BASIN

Plants Many species shared with Cameroon/Gabon but about 12 per cent of lowland rain forest flora thought to be endemic to main part of Zaïre basin (i.e. 'Congolia').

Typical tree species are *Oxystigma oxyphyllum* and *Scorodophloeus zenkeri.*

Invertebrates Numerous insects.

Amphibians Numerous frogs including: *Hemisus olivaceus, Afrixalus osoroi, Afrixalus equatorialis.*

Reptiles *Chameleo quadricornum, Chameleo montium, Chameleo pfeifera.*

Birds Endemics include: Zaïre Dioch *Ploceus anomala,* Purple-banded Sunbird *Nectarinia congensis,* Congo Peacock *Afropavo congensis,* Grant's Bluebill *Spermophaga poliogenys.*

Mammals Endemics include: Schaller's Shrew *Myosorex schalleri,* Swamp Guenon *Cercopithecus nigroviridis,* Salongo Guenon *Cercopithecus salongo,* Wolf's Guenon *Cercopithecus wolfi,* Hamlyn's Guenon *Cercopithecus hamlyni,* Pigmy Chimpanzee *Pan paniscus,* Snouted Mongoose *Xenogale microdon,* Fishing Genet *Osbornictis piscivora,* Giant Genet *Genetta victoriae,* Okapi *Okapia johnstoni.*

Conservation Strategies and Needs in African Countries

The preparation and publication of a conservation strategy for each Nation is a crucial step in rationalizing conservation. Most nations are being lobbied to prepare such plans with assistance from IUCN.

Angola
Some areas already represented in existing reserves (see map) but Angolan escarpment and upland inadequately protected. Civil war has made conservation institutions and action ineffective.
Some needs: To gazette reserves on Angola escarpment ie. Amboim/Gabela. Reserves in uplands ie. Mt Moco. Publish a National Conservation Strategy. Upgrading for Directorate of Nature Conservation. Education in local ecology at all levels.

Cameroon
Most areas already represented in existing reserves (see map) but relatively ineffective due to poorly funded and inadequately manned conservation system which is given too low a priority by Government.
Some needs: To gazette already proposed Forest National Parks, ie. Mbam-Djerem, Dja, Boumba Bek, Nki, Lobéké and Nyong River. To make protection effective in highland areas ie. Etinde and Bambuko Reserves (on Mt Cameroon), Mt Kupe, Mt Oku and other Bamenda Highland Forest Reserves. Publish a National Conservation Strategy. Upgrading for Conservation Department. Education in local ecology. Expand biological research programmes relevant to the management of reserves. Expand education offered at Garoua Wildlife School.

Comoros
Mt Karthala Forest Reserve proposed but forest areas currently being felled without adequate regulation.
Some needs: Publish a National Conservation Strategy Gazette. Make Mt Karthala a Reserve. Education in ecology in schools.

Ethiopia
Most major conservation areas have already been identified and provisional or nominal protection implemented but the Government does not give a high enough priority to the Wildlife Conservation Department and its activities. Bale and Simen Management Plans have been prepared.
Some needs: To gazette the Bale National Park. Implement National Parks management plans. Upgrading for Wildlife Conservation Department. Publish a National Conservation Strategy. Existing programmes of education in ecology to be further developed. Expand biology research programmes relevant to the management of reserves.

Gabon
Large areas of natural forest are undisturbed and already represented in nominal reserves with further reserves proposed (see map). Realistic protection of these reserves currently non-existent except in President's private hunting reserve.
Some needs: To gazette current reserves and extend Wonga-Wongue hunting reserve. Upgrading for Wildlife Conservation Department. Publish a National Conservation Strategy. Education in local ecology at all levels. Expand biological research programmes relevant to the management of reserves.

Ghana

An existing conservation plan is vitiated by population pressure and inadequate protection. Reserves are ineffective due to the poorly funded and inadequately manned Wildlife and National Parks Division of the Forestry Commission being given too low a priority by the Government. Most existing reserves are subject to continuing degradation.

Some needs: Formal gazetting of forest reserves backed up by effective conservation programmes. Upgrading the Wildlife and National Parks Division with a large commitment of personnel, funds and resources. Publish a National Conservation Strategy. Education in local ecology at all levels.

Ivory Coast

Has an extensive system of parks and reserves and hunting has been banned since 1973. Conservation relatively ineffective due to the poorly funded and inadequately manned conservation system and large-scale settlement, poaching of animals and trees and freelance mining, even in the Tai National Park; activity that is sanctioned by powerful interests. West Germany and World Bank are involved in drawing up management plans for major reserves.

Some needs: Upgrading of conservation in national priorities and larger manning and funding of programmes. Publish a National Conservation Strategy. Education in local ecology at all levels. A more forceful protection of Tai National Park from pirate loggers, miners, settlers and hunters. Expand research relevant to the maintenance of the country's unique habitats.

Kenya

One of the most successful wildlife estates in Africa and pioneer in the concept of wildlife clubs in every school in the country. Well maintained parks and reserves in most habitats but Coast (Sokoke and Witu) and Taita inadequately represented, protected and studied. Funds earned from wildlife conservation are often not returned into maintenance, nor shared by those communities that bear the local costs of a tourist economy. The Director of Wildlife Conservation has complained that assistance or support ear-marked for conservation commonly finds its way into 'other unrelated activities'. The consequence of these diversions is that many local people and well-wishers overseas have become alienated from conservation in Kenya.

Some needs: Authorities should follow up on 1974 review of protected areas and implement a higher priority for conservation of endemic communities, ie. Sokoke and Taita. More systematic study and monitoring of Mt Kenya, Aberdare and Elgon upland habitats. More study of year-long environmental needs of rare endemics, notably Grevy's Zebra and Hunter's Hartebeeste; keeping waterpoints open for the former and perhaps closing stock boreholes in central area of hartebeeste range. A more equitable sharing of benefits from tourism and more investment in and extension of environment education. Large-scale commercial charcoal burning should be outlawed in fragile habitats. Greater attention will be needed to the conservation of closed-drainage inland lakes and various sub-desert communities. The experience of Kenyans, especially in wildlife clubs, education and social aspects of conservation should be made accessible to other countries.

Liberia

One recently declared national park with two more planned for the immediate future signal an important change in Liberia. Conservation is a new concept and comes into direct conflict with a well-established trade in bush-meat. With an expanding population (2 million and 17.4 per sq km) and diminishing forest (18% of the land area, but mostly heavily degraded) the areas in which wild animals can be hunted on a commercial scale is declining rapidly. Uncontrolled timber-felling is also nearly universal. In spite of a military government, policies tend to have limited influence because the country is effectively very decentralized.

Some needs: Showing people the long-term consequences of uncontrolled exploitation of natural products may need to be developed on a region by region basis. Existing educational and social institutions at the regional level are probably the main vehicle for development education in ecology. Upgrading the wildlife parks section and a more vigorous protection of rare and threatened species and of key localities, ie. Loffa Mano, Wanegezi, Cavally, Mt Nimba, Krahn Bassa, Grebo, Gola, Gbe etc. Publication of a National Conservation Strategy. Initiate research programmes in wildlife and habitats relevant to their management.

Malawi

This country has an extensive and representative system of reserves, embracing 11.5% of the country in spite of a density of 60 people per sq km. Soil, water and wildlife conservation are all well developed. Lake Malawi, the Nyika Plateau and Mt Mulanje all have legal protection. Malawi's education and management experience could benefit other areas if links could be developed.
Some needs: Isolated forests at Kikala, Mzuma, Mkhwadzi, Sochi and Tayolo contain endemic species and may require further protection. As one of the richest freshwater habitats in the world, Lake Malawi deserves a greatly intensified research programme and careful monitoring. Extension of existing environmental education and research programmes.

Namibia

Has very extensive reserves but a long-standing unresolved political crisis. Limited scope for competing interests continues to be the best guarantee of ecological stability but fencing has seriously affected the larger migratory animals.
Some needs: Publish a National Conservation Strategy. Education in ecology and formation of wildlife clubs.

Nigeria

Up to the present, savanna communities have attracted most attention from conservation authorities. Areas east of the Cross River have suffered a rapid expansion in oil palm, yam and subsistence farming and in massive commercial hunting for bush-meat. A long established Natural History Society has never succeeded in attracting more than a small elite following. Ambiguities in the power of Federal and State Governments have been exploited to subvert the conservation estate and conservation policies. A wildlife school has been set up at New Bussa and courses in wildlife and ecology are now taught at Ibadan University.
Some needs: To gazette Obudu game reserve as a National Park and give added protection and status to other reserves east of the Cross River. To control armed poaching of timber and bush-meat in the Okumu Forest Reserve and to partially or wholly reverse excisions (see p. 128). Upgrading the status, training and manning of the wildlife section of Forestry Department. Publish National or State Conservation Strategies. Expand biological programmes relevant to the management of reserves. Education in local ecology.

Rwanda

The Volcanoes National Park has had large injections of foreign aid and interest. A programme of education in ecology, wildlife and the environment has been initiated in schools. A population density of 185 people per sq km puts limits on the conservation estate.
Some needs: Extension of the education programme. Gazette the Nyungwa Forest as a major conservation area. Continue detailed monitoring of the remaining mountain Gorillas plus basic research. Consider whether recently excised margins of Volcanoes National Park should be returned in order to make habitat of viable size for the Gorilla population.

Sao Tomé and Príncipe

The two islands have a combined area of 964 sq km and a population of approximately 100,000. They are the object of World Bank plantation schemes. Formally recognised reserves have not been declared and are a top priority.
Some needs: To survey endemic flora and fauna especially in the Pico, Portunate and Lagoa Amelía areas of Sao Tomé. Publish a National Conservation Strategy. Gazette national parks and reserves. Education programme in ecology and local biology.

Seychelles

Seychelles has nominated two areas as World Heritage Sites under the UNESCO international convention on the protection of world cultural and natural heritages, namely the Aldabra atoll and the Vallée de Mai on Praslin Island. Nearly 40% of the islands' surface have some protection as nature reserves.
Some needs: Maintain long-term research and monitoring of endemic communities and species such as the Aldabra tortoise population. Also control or exclude exotics in reserves. Expand education in ecology and local biology.

Sierra Leone

The remaining wildlife areas and forests are now very small and subject to intense hunting and timber felling. Protected areas have been declared and more are being considered (see map) but conservation is still low on Government priorities.
Some needs: Gazette Gola Forest (W & E) as a National Park and revoke timber concessions. Institution of a well-funded and effective National Parks and Wildlife Department. Upgrading of conservation in Government priorities. Education in ecology and local biology. Expand research programmes relevant to the management of reserves and parks.

Somalia

Conservation can hardly be expected to be any part of the Somalis immediate preoccupations, which have included territorial conflicts (the war in Ogaden), periodic droughts and famines and now a civil war. None the less a number of Reserves and National Parks exist on paper. Although none of these areas has effective protection or effective official support. The principal body concerned with the environment in Somalia today is the National Range Agency and the most recent recommendations on Somali conservation and management are those of Simonetta and Simonetta (1983).
Some needs: Once younger and more dynamic leadership is forthcoming it is to be hoped that the key status and position of Somalia in pan African and global biogeography will be recognized and acted upon. This would involve a substantial upgrading of conservation and the equipping and funding of an effective wildlife conservation department. Simonetta's recommendations should provide the basis for a conference to draw up a plan of action. Publish a National Conservation Strategy. Introduce education on ecology and the environment into schools. Priority areas for conservation of endemics: Las Anod, Hobyo (Obbia), Jowhar Wrashek, Daalo Forest, Gaan Libax, Ras Guba, El Hammure, El Chebet and Haradere-Awale.

Socotra *(Revolutionary Republic of South Yemen territory)*

As yet no protected areas. Northern parts of the main island and parts of Abd el Kuri are priority areas for conservation, ie. Jebel Hamadar and Jebel Haggier.
Some needs: To alert the appropriate Yemeni authorities to global interest in this unique Centre of Endemism. International support for a conservation programme.

South Africa

The highly diversified nature of conservation in South Africa is its most unusual feature. There is private, public and institutional involvement in conservation, education and research at all levels but reaching a limited spectrum of the public. In the longer term reserves will need to be projected as universal assets. At present some fynbos and Karroo types are not adequately represented in conservation areas but the number and distribution of reserves in the Cape are impressive (see map). The country publishes its own Red Data Books on endangered species.
Some needs: Larger botanical reserves in key areas: Caledon, Worcester and lowland fynbos zones. Extension of ecological education, natural history clubs and expeditions to wilderness areas to a wider section of the urban public. Enlarge scope for training (and exchange) in environmental conservation and management.

Sudan

The Sudd swamps are the only habitat for many rare animals and plants. Unless or until plans for the Jonglei Canal are resurrected this vast area is not susceptible to any form of intervention. Legal protection of the Nile Lechwe, Shoebill and other species needs to be maintained but is unlikely during the current civil war.

Tanzania

Tanzania has made very substantial efforts to conserve its natural resources and has very large areas under some form of nominal control or protection but has inadequate practical protection in its key areas of local endemism. Tanzania has its own network of wildlife clubs known as the Malihai clubs which actively promote education in the environment and conservation.

Some needs: Review and consolidate and give greater priority and meaningful protection to long-term reserves in the Usambaras and Ulugurus. Gazette the Mwanihana National Park and initiate conservation of Chita Forest and south Zanzibar as future National Parks. Better protection for the Pugu Hills Forest and Rondo. Expand research programmes relevant to the management of reserves, especially those in Centres of Endemism. Extend education programmes in ecology, the environment and conservation at all levels. Maintain ambitious, broad-based programmes of education at Mweka Wildlife College and expand to include awareness of endemic flora and fauna.

Uganda

Uganda has an extensive system of parks and reserves and a long-deferred plan to set up eight new national parks by upgrading existing conservation areas. Fifteen years of economic and military mismanagement (a local euphemism) have undermined most institutions and estates such as game and forest reserves. General Idi Amin damaged Uganda's conservation estate in two ways. He allocated pieces of reserves to cronies or as rewards. He also flooded the country with dispossessed peasants who were disinherited due to his Government's selling land titles to individual applicants for traditionally held, group-owned or 'Mailo' land. These unfortunate people now 'squat' in forest and other reserves and provide a major challenge for any effort at rehabilitation. In spite of these and other problems and immense economic difficulties there is today a cadre of educated, vigorous and forward-looking people with strong commitments to conservation of the environment and wildlife. UNDP and EEC are currently helping to rehabilitate the national parks.

Some needs: Rehabilitating the malfunction of an established system of parks and reserves is the first priority but effective conservation of the Impenetrable and other West Uganda Forests is also in hand and needs international support, especially in relation to environmental education and management oriented research. Gazetting of the Impenetrable Forest as a National Park and the Ruwenzori Mountains as World Heritage Sites. Re-orientating of the Forest Department to make conservation and research major rather than subsidiary parts of their function should be a specific component of the National Conservation Strategy currently under preparation. Amin's excisions from the Echuya Forest and swamp (key habitats in Kigezi) should be revoked. Also areas of the gorilla sanctuary on Mts Muhavura and Sabinio should be returned to reserve status. The wildlife clubs of Uganda and other initiatives in conservation education need expansion and support. Basic ecological research through Makerere University and the Uganda Institute of Ecology needs to be expanded and given adequate support.

Zaïre

The extent of Zaïre, 2.3 million sq km (nearly half forest), is some guarantee for the survival of most species but the rapidly growing population on the eastern margins of Zaïre threatens many of the montane habitats where there is a great concentration of endemics. Although conservation is given a high priority by the government there is still inadequate provision for effective protection of some reserves.

Some needs: Establish new reserves in Lomako (mainly for Pygmy Chimpanzees), Uvira, Marungu and in the Itombwe Mountains. Establish country wide wildlife clubs and education in local ecology. Expand biological research programmes. Publish a National Conservation Strategy.

World Heritage Sites and Biosphere Reserves in African Countries with Major Centres of Endemism

Reserves in italics are those within Centres of Endemism

WORLD HERITAGE SITES

Ethiopia
Simen National Park

Guinea
Mount Nimba Strict Nature Reserve

Ivory Coast
Comoe National Park
Tai National Park
Mount Nimba Strict Nature Reserve

Malawi
Lake Malawi National Park

Seychelles
Aldabra Atoll
Vallée de Mai Nature Reserve

Tanzania
Ngorongoro Conservation Area
Serengeti National Park
Selous Game Reserve

Zaïre
Kahuzi-Biega National Park
Salonga National Park
Virunga National Park
Garamba National Park

BIOSPHERE RESERVES

Cameroon
Dja Reserve
Benoue National Park
Waza National Park

Gabon
Ipassa Makokou Reserve

Ghana
Bia National Park

Guinea
Massif du Ziama Biosphere Reserve
Mount Nimba Biosphere Reserve

Ivory Coast
Comoe National Park
Tai National Park

Kenya
Mount Kulal Biosphere Reserve
Mount Kenya Biosphere Reserve
Kiunga Marine National Reserve
Malindi-Watamu Biosphere Reserve

Nigeria
Omo Reserve

Rwanda
Volcanoes National Park

Sudan
Radom National Park
Dinder National Park

Tanzania
Serengeti N.P. & Ngorongoro C.A.
Lake Manyara National Park

Uganda
Queen Elizabeth National Park

Zaïre
Yangambi Floristic Reserve
Luki Floristic Reserve
Lufira Valley

African Reserves and National Parks

Seychelles
Proposed National Park—Praslin N.P.
Vallée de Mai Nature Reserve

Aldabra
Aldabra Biosphere Reserve

Zanzibar
Jozaní designated sanctuary, Uzi Forest Reserve

St Helena
High Peaks Forest Reserve

Republic of South Africa
NATIONAL PARKS
Golden Gate Highlands NP, Royal Natal NP, Mountain
Zebra NP, Addo Elephant NP, Tsitsikama Forest and Coast
NP, The Bontebok NP.

GAME, NATURE AND FOREST RESERVES
Ndumu GR (birds), Kosi Bay NR, Umfolozi GR, Nkundla
FR, Mkuzi GR, Hluhluwe GR, Dukudula FR, Himerville
NR, Coleford NR, Umlalozi NR, Richards Bay GR, Oribi
Gorge NR, Giant's Castle GR, Quinafalls FR, Dlinza FR,
Rust de Winter NR, Melville Koppies NR (indigenous
flora), Krugersdorp GR, Lotani NR, Coleford NR,
Kronskloof NR, Vernon Crooks NR, Willem Pretorius GR,
Settlers Park NR (indigenous flora), Silvermine NR, Tokai
FR, The Flora Reserve and Duckitt's Farm, Cape of Good
Hope NR, Betty's Bay (Harold Porter Botanical Reserve).

MISCELLANEOUS
Knysna Forest, Cedarberg WA, Rondevlei Bird Sanctuary,
Kirstenbosch National Botanical Gardens, Newlands Forest.

**Namibia (South West Africa) and the Karoo (within the
Republic of South Africa)**
NATIONAL PARKS
Augrabies Falls NP, Etosha NP, Namib-Naukluft NP.

GAME, NATURE AND FOREST RESERVES
Daan Viljoen GP, Nieuwoudtville NR (indigenous flora),
Carnarvon NR, Ramskop NR, Hester Malan NR,
Strydenburg Aalwynprag NR, Fish River NR, Tsaobis-
Leopard NR, Omaruru NR, National West Coast TA.

MISCELLANEOUS
Ai-Ais Hot Springs, Waterberg Plateau Park, Petrified
Forest (*Welwitschia mirabilis*).

Somalia
NATIONAL PARKS
Lack Dere NP, Lack Badana NP, Daalo Forest NP, Gaan
Libax NP, Rus Guba NP, Jowhar-Warshek NP, Arbowerow
NP, Angole-Farbiddu NP.
    Proposed National Parks—Las Anod, Taleh-El-Chebet.

WILDLIFE RESERVES
Boja Swamps WR, Far Wamo WR, Hobyo WR, El
Hammure WR.

Ethiopia
NATIONAL PARKS
Awash NP, Simien NP, Omo NP.
    Proposed National Parks—Abiyata-Shalla Lakes, Nechsar,
Bale Mountains, Yangudi Rassa.

WILDLIFE SANCTUARIES AND RESERVES
Yabelo, Ever, Awash West, Shire, Harrar WA, Gewane WR,
Tama WR, Chew Bahar WR.

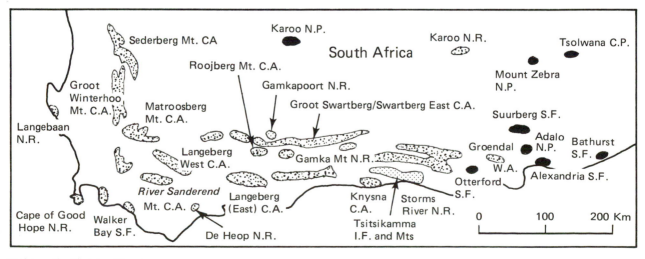

Parks on the Skeleton Coast.

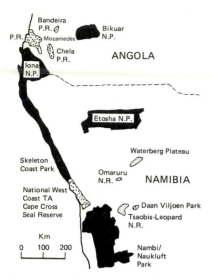

Parks on the Skeleton Coast.

CONTROLLED HUNTING AND CONSERVATION AREAS
Borena, Arsi, Boye, Segen, Gambella CA.

Kenya
NATIONAL PARKS
Aberdare NP, Mt Kenya NP, Gedi NP, Mt Elgon NP (severely disturbed), Sibiloi NP.

NATIONAL RESERVES
Shimba Hills (proposed National Park), Sokoke-Arabuko FR, Mts Nyiru, Ndoto FR, Aberdare Kinangop FR, Tana River FR (severely disturbed), Witu, Boni, Dodori FR (severely disturbed).

Tanzania
NATIONAL PARKS
Mikumi NP, Mt Kilimanjaro NP, Arusha and Ndurgoto Crater NP, Mahali Mountains NP.
    Proposed National Parks—Chita FR, Mwanihana FR.

NATIONAL RESERVES
Mt Meru GR, Umalila-Poroto Ridge (Ishinga and Lumwe valley) FR, Rungwe Mountain FR, Livingstone–Kipengere Complex FR, Njirikiru, Luponde, Silupati FR, Sakarangumo, Madenge FR, Uzungwa–Kilongbero Scarp Complex FR,

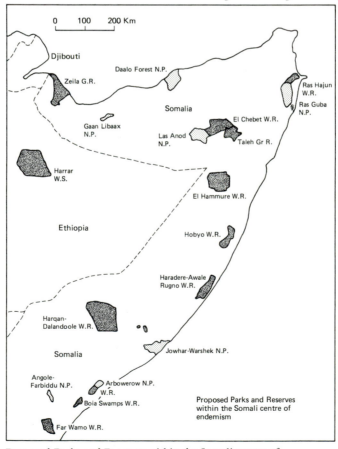

Proposed Parks and Reserves within the Somali centre of endemism.

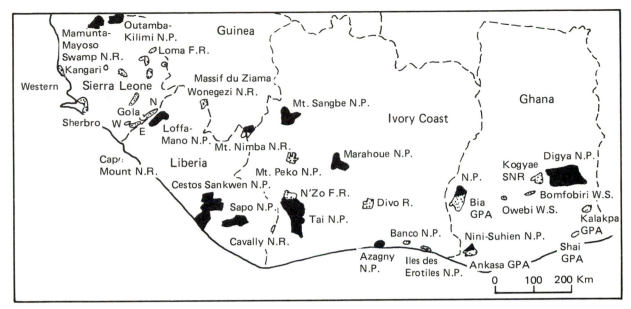

Parks and Reserves within the Upper Guinea centre of endemism.

Uzungwa FR, Ihange FR, Ikwambi Dabaga FR (severely disturbed), Magombera FR, Matengo Hills FR, Rondo FR, Ulugurus North and South FR (severely disturbed), North and South Ngurus FR, Eastern Usambar as FR (severely encroached), Western Usambaras FR (severely encroached).

Uganda
NATIONAL PARKS
Ruwenzori NP (Kabalega), Murchison Falls NP, Queen Elizabeth NP.

    Proposed National Parks—Bwindi (Impenetrable) FR (disturbed), Kibale FR (distubred), Mt Elgon FR (severely disturbed), Ruwenzori Mountains FR (disturbed), Kibale Corridor (severely encroached).

NATIONAL RESERVES
Bufumbira Gorilla Sanctuary (severely disturbed), Budongo Siba FR (disturbed), Bwamba FR, Sango Bay FR, Maramagambo–Kalinzu FR, Kashohya–Kitomi FR (severely encroached), Echuya FR (severely encroached).

Sierra Leone
Proposed National Park—Outamba–Kilimi.

NATIONAL RESERVES
Gola FR, Malay FR, Kambui Hills FR.

Liberia
NATIONAL PARKS
Sapo NP, Loffa Mano NP, Cestos Senkwen NP, Mount Nimba NP.

NATIONAL RESERVES
Wonegezi NR, Cape Mount NR, Cavally NR, Mount Nimba NR.

Ivory Coast
NATIONAL PARKS
Tai NP (severely disturbed), Banco NP (a plantation, severely disturbed), Marahoue NP (threatened by commercial interests).

NATIONAL RESERVES
N'zo FR, Divo Reserve.

Ghana
NATIONAL PARKS
Nini-suhien NP, Diagya NP, Bui NP.

FOREST RESERVES
Bia Tawya, Ankasa, Krokusua (severely encroached).

Cameroon
NATIONAL PARKS
Korup NP.

    Proposed National Parks—Mbam Djerem, Dja River FR.

NATIONAL FOREST RESERVES
Takamanda FR, Fungom FR, Nia Ali FR, Bakala FR, Mungo River FR, Rumpi Hills FR, Meme River FR, Mokoko FR, Bambuko FR, Bakendu FR, Dibombe Mabobe FR.

    Proposed National Reserves—Mount Koupe, Mount Oku (seriously damaged).

COUNCIL FOREST RESERVES
Banyang Mbo FR, Ejagam FR, Bali–Ngemba FR, Bafut Ngemba FR.

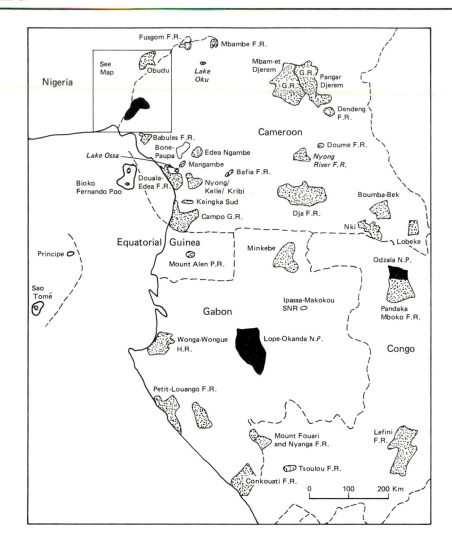

National Parks and Reserves in the Cameroon/Gabon centre of endemism.

Nigeria (South East)
Proposed National Park—Obudu-Boshi-Okwango NP.

DEGAZETTED OR UNCERTAIN
Rivers GR Sections A & B, Cross River GR Sections A & B, Stubbs Creek GR, Anambra GR.

Angola
NATIONAL PARKS
Iona NP, Bikuar NP, Mupa NP, Kameia NP, Kisama NP.

Zaïre
NATIONAL PARKS
Maiko NP, Kahuzi–Biega NP, Park des Virungas, National Park de Salongo.
    Proposed National Park—Ituri Okapi NP.

GAME NATURE AND FOREST RESERVES
Shaba GR, Lomako Yekokora GR, Yangambi FR, Bombo-Lumene GR, Bomu NR.

Rwanda
NATIONAL PARKS
Volcans NP (Gorillas), Akagera NP.

Gabon
NATIONAL PARK
Lupe–Okanda NP.

FOREST RESERVES
Petit–Louango FR, Moukaluba FR, Minkebe FR.

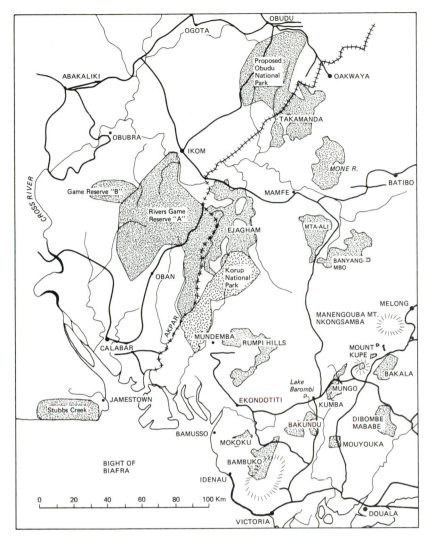

National Parks and Reserves east of the *Cross River* (Nigeria and Cameroon).

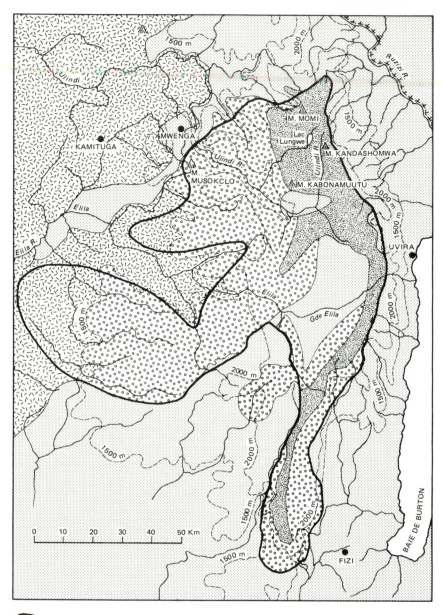

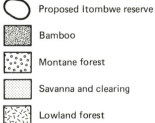

○ Proposed Itombwe reserve

▓ Bamboo

○○ Montane forest

░ Savanna and clearing

⫶ Lowland forest

The Itombwe Mts

African Natural History Societies and Clubs and Other Organizations with African Interests

 East African Natural History Society. P.O. Box 48019, Nairobi, Kenya.

Kenya Wildlife Clubs, c/o National Museum, Nairobi, Kenya.

Southern Africa Nature Foundation, P.O. Box 456, Stellenbosch 7600.

 Somali Ecological Society c/o Nodlaig Guinan, P.O. Box 1789, Mogadishu, Somalia.

 Impenetrable Forest Conservation Project, Zoology Department Makerere University, P.O. Box 7062, Kampala, Uganda.

 (Uganda Wildlife Clubs), The Uganda Society, P.O. Box 4980, Kampala, Uganda.

 The Tanzanian Society, P.O. Box 511, Dar-es-Salaam, Tanzania.

 Wildlife Conservation Society of Zambia, P.O. Box 30255, Lusaka, Zambia.

 The Zoological Society of Southern Africa.

 Botanical Society of S. Africa, Kirstenbosch, Private Bag X7,

Claremont 7735, Cape Province, South Africa.

Namibia, Southwest Africa Clanwilliam Wild Flower Garden, P.O. Box 24, Clanwilliam 8135.

 Ivory Coast, Cote d'Ivoire Nature, Boite postale 1776, Abidjan 08.

 Nigerian Conservation Foundation, Mainland Hotel, P.O. Box 467, Lagos, Nigeria.

 The Wildlife Conservation Society of Tanzania, Dar-es-Salaam.

 Ethiopian Wildlife and Natural History Society, P.O. Box 60074, Addis Ababa.

Wildlife Clubs of South Sudan, Sabra, c/o Zoology Department Juba University, Juba, South Sudan.

The Gambian Ornithological Society, P.O. Box 757, Banjul, Gambia.

The West African Ornithological Society (publish *Malimbus*). I. Fishers Heron, East Mills, Fordingbridge, Hants., FP6 2JR.

Zambia Ornithological Society, Box 33944, Lusaka, Zambia.

Malawi National Fauna Preservation Society (publishes *Nyala*), Box 30370, Lilongwe, Malawi.

Ornithological Association of Zimbabwe (publish *Honeyguide*), Box 8382, Causeway, Harare, Zimbabwe.

Botswana Bird Club (publish *Babbler*), P.O. Box 71, Gabarone, Botswana.

The Cameroon Forest Parks Conservation Education Programme c/o PAMOL PMB 55 Victoria Cameroon. (Publishes 'Sumbu').

SWA/Namibia Bird Club, Box 67, Windhoek 9000.

S.A. Ornithological Society (publish *Bokmakierie* and *Ostrich*), Box 87234, Houghton, Joburg 2041.

Cape Bird Club (newsletter *Promerops*) P.O. Box 5022, Cape Town 8000.

Percy Fitzpatrick Institute University of Cape Town, Rondebosch 7700 SA (publish *Cormorant*).

CONSERVATION SOCIETIES

IUCN, International Union for the Conservation of Nature (has subsidiaries among them, The Species Survival Commission, SSC, and the Conservation Monitoring Centre CMC.), Morgas, Switzerland

FAUNA & FLORA
PRESERVATION SOCIETY
Conserving Wildlife Since 1903

Fauna and Flora Preservation Society, 79–83 North Street, Brighton, East Sussex BN1 1ZA U.K.

Missouri Botanical Gardens, P.O. Box 299, Saint Louis, Missouri 63166 USA.

World Wide Fund for Nature WWF, 1196 Gland, Switzerland

National Geographic Society 17th 7 M Streets, Washington DC 20036, USA

Greenpeace

EARTHSCAN, International Institute for Environment and Development

The Royal Geographical Society, 1 Kensington Gore, London

African Wildlife Foundation, 1717 Massachusetts Avenue NW, Washington DC 20036, USA.

New York Zoological Society, New York, USA.

Friends of the Earth, 377 City Road, London EC1V 1NA, U.K.

Frankfurt Zoological Society.

International Board for Plant Genetic Resources, Royal Botanic Gardens, Kew, Richmond, Surrey, U.K.

Common Ground International, 25 Downshire Hill, Hampstead, London NW3 1NT, U.K.

Royal Society for the Protection of Birds, The Lodge, Sandy, Beds., SG19 2DL, U.K.

International Council for Bird Preservation, 32 Cambridge Road, Girton, Cambridge, CB3 0PJ, U.K.

Jersey Wildlife Preservation Trust

Peoples' Trust for Endangered Species

World Forest Action

Worldwatch Institute

International Centre for Conservation Education

The WWB, Fund for International Bird Conservation, 3030 Walsrode, West Germany.

Index

Species and places mentioned in the appendices, being easy to refer to, are not listed here. Figures in *italics* denote page numbers of captions to illustrations, frequently with textual mention on same page.